THE SUSTAINABILITY AND CLIMATE CHANGE CURRICULUM OUTDOORS

KEY STAGE 2

Alun Morgan, Deborah Lambert, Michelle Roberts, Sue Waite

BLOOMSBURY

BLOOMSBURY EDUCATION
Bloomsbury Publishing Plc
50 Bedford Square, London, WC1B 3DP, UK
29 Earlsfort Terrace, Dublin 2, Ireland

BLOOMSBURY, BLOOMSBURY EDUCATION and the Diana logo are trademarks of Bloomsbury Publishing Plc

First published in Great Britain, 2024 by Bloomsbury Publishing Plc

Text copyright © Alun Morgan, Deborah Lambert, Michelle Roberts, Sue Waite, 2024

Illustrations copyright: Figure 1.1 © United Nations (n.d.); Figure 1.3 © IPCC (2014);
Figure 1.4 © The Children & Nature Network (n.d.); Figure 1.5 © Paul Warwick (2017);
Figure 1.6 © Hannele Cantell (2019); Figure 2.1 © Michelle Roberts (2022)

Photographs © Alun Morgan, Deborah Lambert, Michelle Roberts, Sue Waite, 2024

Material from Department for Education documents used in this publication
are approved under an Open Government Licence:
www.nationalarchives.gov.uk/doc/ open-government-licence/version/3/

Alun Morgan, Deborah Lambert, Michelle Roberts, Sue Waite have asserted their rights under the
Copyright, Designs and Patents Act, 1988, to be identified as Authors of this work

A catalogue record for this book is available from the British Library

ISBN: PB: 978-1-8019-9275-6; ePDF: 978-1-8019-9274-9; ePub: 978-1-8019-9272-5

2 4 6 8 10 9 7 5 3 1 (paperback)

Text design by Marcus Duck
Typeset by Marcus Duck

Printed and bound in the UK by CPI Group (UK) Ltd., Croydon, CR0 4YY

To find out more about our authors and books visit www.bloomsbury.com
and sign up for our newsletters

Contents

List of abbreviations

CK	Content Knowledge
COP	Conference of the Parties
CPD	Continuing professional development
CPPD	Continuing professional and personal development
DEFRA	Department for Environment, Food and Rural Affairs
DfE	Department for Education
INSET	In-service training
KS	Key Stage
LKS2	Lower Key Stage 2
n.d.	no date
OPK	Outdoor Pedagogical Knowledge
PCK	Pedagogical Content Knowledge
PK	Pedagogical Knowledge
RAG	Red, amber, green (rating)
SCCB	Sustainability climate change and biodiversity
SCCBCK	Sustainability climate change and biodiversity Content Knowledge
SCCS	Sustainability and Climate Change Strategy
SDGs	Sustainable Development Goals
SEMH	Social, emotional and mental health
UKS2	Upper Key Stage 2
UNESCO	United Nations Educational, Scientific and Cultural Organization

Acknowledgements

We would like to thank the many young people who have urged action on sustainability and climate change across the world. This book is written in response to their demands for greater recognition of the importance of children's education. We would like to thank the Wild Tribe team for their support in developing the 'Earth Tribe' programme, which has supported the piloting of some of the lessons.

Foreword

The publication of this book comes at a time of an ever-increasing global narrative that society needs to urgently transition towards more sustainable futures, particularly in the light of the latest research on climate change (UNESCO, 2015; IPCC, 2021). This transformative learning agenda is calling for the education sector as a whole to find new ways to engage all citizens to make vital contributions to this transition. An example of this move within the context of England is the Department for Education's Sustainability and Climate Change Strategy, which has provided the context for this publication (Department for Education, 2022). This strategy rightly emphasises the need for educational approaches that enhance and enable nature connection and for this to be readily available and on the doorstep of every child. I would argue that this connection should be established in several ways. It should capture a playful curiosity and appreciation of wonder and awe. Additionally, it should be informative about the nature of the threats posed to the environment and our own wellbeing. It should also highlight the opportunities we have for making positive, sustainable changes together. This strategy also rightly highlights the vital role that is played by primary schools in the formative years of our children's development. This context feeds into why I am so hugely encouraged by the theme and contents of this publication. With its focus on taking the curriculum outdoors at Key Stage 2, it is just the source of inspiration and practical support needed at this time within the education sector.

This book was written by a collection of deeply committed professionals who command well-established reputations as academic researchers and practitioners in this field. This has resulted in an engaging and uplifting book that follows the genetic code of what, I would argue, makes for apt sustainable education, as it engages this crisis with the necessary urgency but also with a spirit of hope and opportunity. I very much hope this publication makes a vital contribution to an ever-growing community of practice that seeks to revitalise education for the common good. In so doing, I have every confidence that it will support a generation of young learners who are able to face the challenges of their times with ingenuity, resilience and connection. This will enable them to playfully and imaginatively engage in place-making with their own landscapes for life.

Dr. Paul Warwick
Centre for Sustainable Futures Lead, Plymouth Institute of Education, University of Plymouth

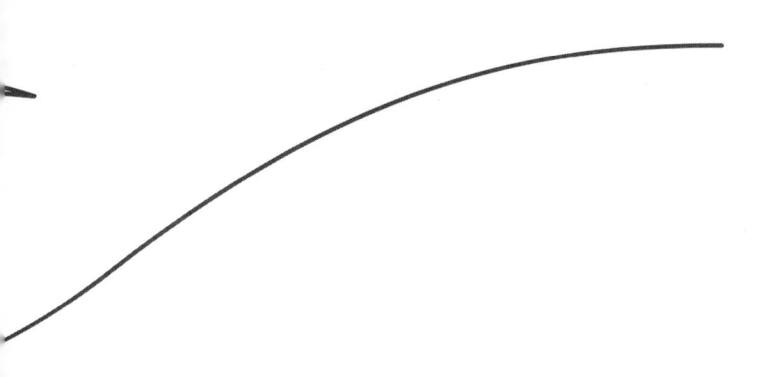

Introduction

This book has emerged in response to the UK government's Sustainability and Climate Change Strategy or SCCS (Department for Education, 2022), and relevant extracts from this document can be seen in Appendix 1. This marked an exciting moment for the school community to once again focus attention on some of the most pressing 21st-century challenges facing humanity, namely sustainability, climate change and biodiversity (SCCB) loss. This book is intended to support teachers to creatively engage with these issues in the context of the strategy, with a particular focus on using the outdoors as a crucial learning context to achieve its aims. A justification for this perspective is presented in Chapter 1, with the remainder of the book providing advice, guidance and exemplary material to empower teachers to either embark or improve on their own creative and fulfilling professional journey as outdoor practitioners of Environmental and Sustainability Education.

Each chapter attempts to support teachers, individually and collectively, in developing their Pedagogical Content Knowledge (PCK) in relation to specific topics, educational orientations and learning contexts. Shulman (1987) originally conceptualised the idea of PCK to discuss the type of knowledge that is unique to teachers, which combines Content Knowledge (of a subject or specific topic) with Pedagogical Knowledge (general principles of how to teach) to arrive at the best pedagogical approach to address a particular aspect of content to specific learners in a particular context (see Figure 0.1).

It is safe to assume that teachers have good Content Knowledge (CK) in relation to the specific National Curriculum subjects they teach. However, SCCB is likely to feature new content that teachers might need to grapple with before feeling confident teaching it. Similarly, it is safe to presume that teachers have good Pedagogical Knowledge (PK) as classroom practitioners through their initial and in-service training (or continuing professional development), and through their ongoing reflective practice in the classroom. What perhaps requires further development are those aspects of PK specifically associated with learning outdoors (such as health and safety, group management, etc.). This can be thought of as Outdoor Pedagogical Knowledge (OPK). Research has shown that the lack of OPK and associated confidence in relation to outdoor learning represents a key barrier to taking lessons and learning outside the classroom. When combined with a lack of familiarity with the specific SCCB content, this barrier is compounded. Consequently, this book intends to provide teachers with suggestions that will inspire them to develop their own CK, OPK and PCK in relation to teaching the intrinsically connected SCCB.

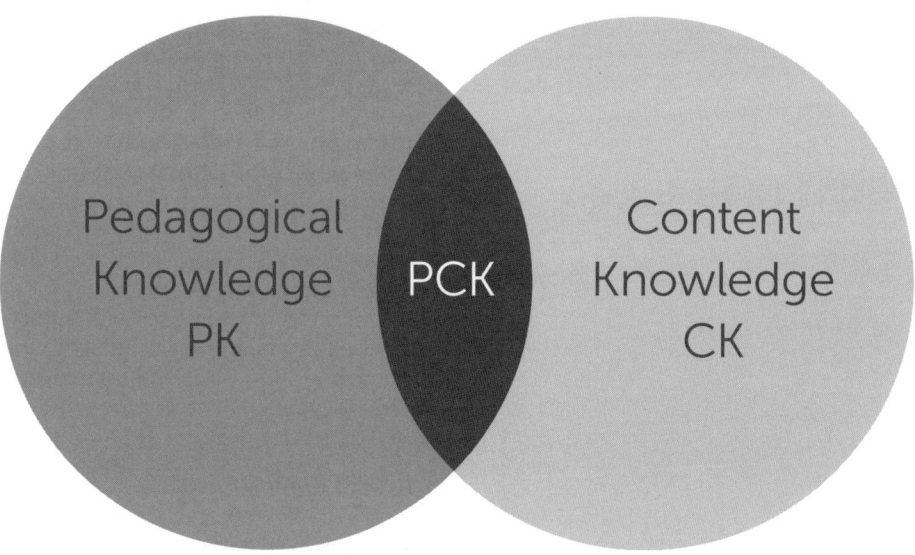

Figure 0.1: Pedagogical Content Knowledge (PCK) (based on Shulman, 1987).

How to use the book – chapter-by-chapter

Chapter 1 provides basic background knowledge on the topics of SCCB.

Chapter 2 discusses the role of leadership in effecting sustainable change, providing exemplar models.

Chapter 3 considers ways to develop the natural infrastructure within and beyond school grounds, including case study examples of good practice and boxed exercises to help you reflect on your own practice.

Chapter 4 presents a set of six exemplar progressions for Years 3 and 4 (lower Key Stage 2) and **Chapter 5** presents a set of six exemplar progressions for Years 5 and 6 (upper Key Stage 2). These show how knowledge and skills can be progressively developed across the Key Stages to include SCCB. While these do prioritise science, they are intended to integrate other subject material in a cross-curricular manner. They are also intended to incorporate opportunities to develop broader, 21st-century skills and competencies such as communication, personal and social development and citizenship.

Chapter 6 briefly reviews the book and invites readers to creatively continue their engagement in this vital work in the future.

Photos and further resources

Illustrative photos are available online at bloomsbury.pub/sustainability-curriculum-outdoors. When you see references to the Bloomsbury Education website or this online logo: ◑, it means that the resources are available to download from the URL above. Further resources are also available at the back of the book in the Appendices.

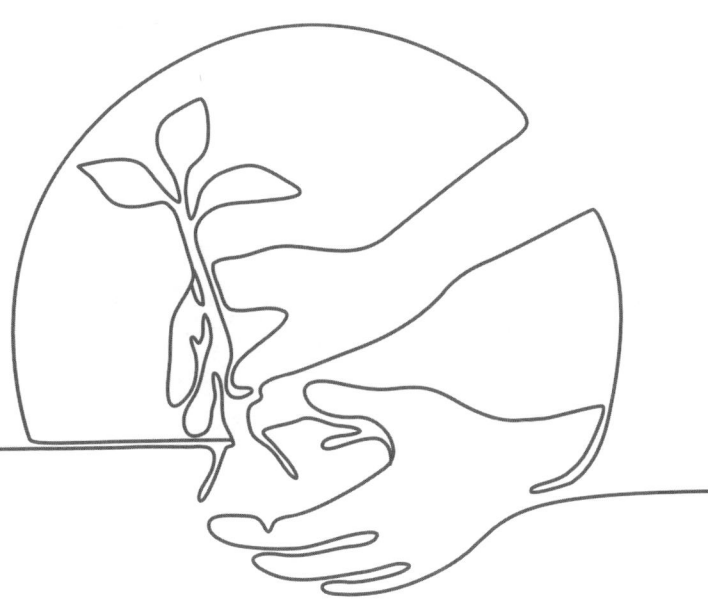

Chapter 1
Sustainability and climate change in upper primary – a rationale

A turning point towards hope

This book is oriented around a 'pedagogy of hope', looking positively and creatively to identify sustainable solutions at the school level that can contribute to sustainability beyond the school grounds. The real beauty of such an approach is that it not only makes an invaluable contribution to sustainability locally (and ultimately globally), but it simultaneously improves the quality of learning, health and wellbeing of pupils (and teachers), as described throughout the book.

With the UK government's Sustainability and Climate Change Strategy or SCCS (DfE, 2022), real potential arose for schools to contribute to and benefit from these multiple and mutually reinforcing positive outcomes. The strategy was a key impetus behind this book, as reflected in the title. The book is, therefore, intended to support teachers in primary schools with their implementation of the strategy through appropriate educational response to the challenges we face.

The contemporary world can be a troubling place for both teachers and young people alike. Apart from the day-to-day performance and accountability pressures (Ofsted inspections, performance criteria, test scores, etc.) and everyday life (financial pressures, interpersonal relations, etc.), we are increasingly bombarded with information about serious, wide-reaching issues. These issues affect all of us as members of our local communities and an increasingly global society. They can negatively affect our quality of life and, in other words, affect our sustainability. Here, sustainability refers to our capacity to prosper as individuals and communities at all levels (locally to globally) now and in the future.

The notion of sustainability has been around for several decades and has been adopted in everyday language. However, the concept is quite complex, incorporating a wide range of interconnected themes and issues. Understanding these, along with climate change and biodiversity loss, are key foci of this book and outlined in greater detail below. All three issues are becoming so serious that people have started to talk of them as having reached crisis levels.

Threats to sustainability – whether social, economic or environmental (which are often too interconnected to neatly separate) – can be quite worrying and anxiety-inducing, especially when we refer to them as 'crises' or frame them in terms of 'doom and gloom'. This is counterproductive, as it causes people to feel overwhelmed and powerless to deal with issues apparently beyond their ability to affect, leading to inaction. The good news is that everyone has the potential to make a positive change by taking appropriate action at a scale that they can influence. This is encapsulated by the phrase 'Think Global, Act Local'.

Everyone, including teachers and students, has the potential to become part of the solution by becoming empowered 'change agents' through developing the necessary competencies (knowledge, skills, attitudes and values) to work collaboratively in their home locality to make the world a better and more sustainable place.

Sustainability explained in a nutshell

According to the famous Brundtland Report, sustainability (or sustainable development) is about 'meeting the needs of the present generation without compromising the ability of future generations to meet their needs' (World Commission on Environment and Development, 1987). The idea stresses the interrelationship between three key domains or pillars of human existence, which must be in balance to achieve sustainability:

• social

• economic

• environmental.

These three pillars need equal consideration to achieve sustainability. The concept of equal consideration can be explained through an analogy with a three-legged stool. For a three-legged stool to function, all legs must be of equal importance. If any leg is missing or of a different length, the stool loses its function. It is unbalanced and does not work. In the context of sustainability, the social, economic and environmental pillars must be of equal importance (worth and value) to deliver effective sustainability.

Historically, sustainability was popularised at the Rio Earth Summit (otherwise known as the United Nations Conference on Environment and Development or UNCED) in 1992. This key international conference introduced the idea of 'Agenda 21' (UNESCO, 1992), which explored the activities that humanity must undertake collaboratively to solve the most pressing issues at both local and global levels. Indeed, it was recognised that most of the necessary work must take place at the local and community level, as encapsulated by the phrase, 'Think Global, Act Local'. Education was recognised as having a key role in moving the world community towards sustainability, with Chapter 36 of Agenda 21 being devoted to it. Schools were identified as key partners in achieving sustainability. This is now reinforced in England by the statutory requirement for sustainability and climate change education (DfE, 2022).

The Rio Earth Summit also set up three conventions (known as the Rio Conventions) on the most pressing problems, two of which represent the key foci of this book:

• United Nations Framework Convention on Climate Change (UNFCCC)

• UN Convention on Biological Diversity (UNCBD)

• UN Convention to Combat Desertification (UNCCD).

The Intergovernmental Panel on Climate Change (IPCC) and Intergovernmental Science-Policy Platform on Biodiversity and Ecosystem Services (IPBES) formed expert panels to regularly report on the 'state of the planet'. In 2021, the two expert panels worked together to publish a Joint Report demonstrating the inextricable links between climate change and biodiversity (Pörtner et al., 2021).

There are also regular (typically annual) meetings of representatives of signatories of the conventions which are called Conference of the Parties (COP). These meetings are often in the news, but it is easy to confuse those for climate change with those for biodiversity as they are both called 'COP' followed by a number. The UK Government hosted COP 26 (the UN Climate Change Conference) in Glasgow in 2021, which provided an important driver behind the recent policy shifts, including the SCCS. Conversely, a Biodiversity Conference (COP 15) was held in Montreal in 2022, which was important for presenting a new Global Diversity Framework.

Since the Rio Earth Summit in 1992, the concept of sustainability (or sustainable development) has been developed through a series of summits. In 2015, the UN General Assembly presented a 'shared blueprint for peace and prosperity for people and the planet, now and into the future' to work with until 2030. This interprets sustainability as compromising 17 interrelated Sustainable Development Goals, or SDGs (shown in Figure 1.1).

Figure 1.1 The 17 United Nations Sustainable Development Goals, reprinted with permission from the United Nations (United Nations, n.d.). The content of this publication has not been approved by the United Nations and does not reflect the views of the United Nations or its officials or Member States.

Climate change explained in a nutshell

The Earth's climate is dynamic and subject to changes for a variety of reasons, such as shifts in the Earth's orbit around the Sun, changing behaviour of the Sun in terms of its emission of solar energy, and/or the introduction of chemicals into the atmosphere that affect the amount of incoming and outgoing energy. These changes can be brought about quite naturally and have resulted, over geological timescales, in periods when the Earth has experienced extremes of temperature, such as ice ages or hotter temperatures associated with the Jurassic period.

However, the term climate change usually describes the contemporary rapid increase in global average temperatures, identified by scientists and recognised as human-induced (or anthropogenic) in the last hundred years or so. Activities associated with humanity are understood to be exceeding the natural processes outlined above. Previous terms such as 'global warming' have fallen out of favour, and it is more accurate to speak of 'Anthropogenic Global Climate Change' (i.e. the human-induced global increase in average temperatures).

The 'greenhouse effect' is the phrase used to explain this warming because, metaphorically, it is similar to how a greenhouse works. Solar energy from the Sun arrives from space to the Earth. This is high energy, or shortwave, radiation which can easily pass through our atmosphere (the glass of the greenhouse) to reach the surface of the Earth, where some is absorbed to heat the planet. Some of this energy is, however, 'bounced' back by the Earth's surface (or, technically, re-radiated by the planet). In so doing, the energy is transformed into longer wavelength heat with lower energy. Some of this can pass back out into space, but some is 'trapped' within certain gases in the atmosphere (the glass of the greenhouse), such as carbon dioxide (CO_2), methane (CH_4) and water vapour, which are all naturally occurring. In effect, this warms the atmosphere and planet (inside the greenhouse). Figure 1.2. provides a simple diagram to explain this process.

The greenhouse effect is crucial for the future existence, or sustainability, of life as we know it. Without the warming effect, the Earth would be a frozen, ice planet. The current issue is that the actions of humans over the last couple of hundred years have greatly increased the amount of greenhouse gases in the atmosphere, and consequently exponentially increased the warming process (see Figure 1.3).

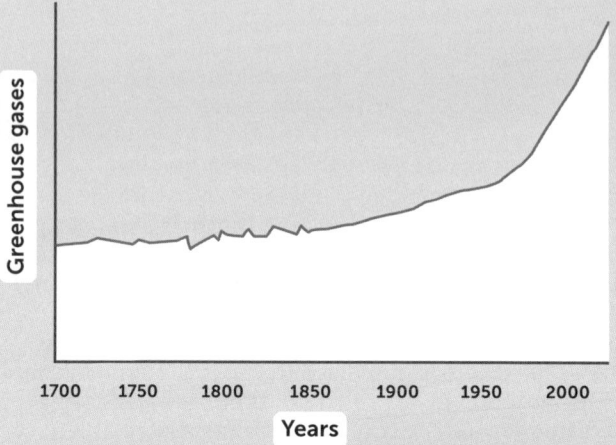

Figure 1.3. The human-induced rise in global warming, the increase in greenhouse gases since c1700.

Our contemporary, energy-hungry, consumer lifestyles are associated with the combustion of fossil fuels (coal, oil and natural gas) which pump out large quantities of carbon dioxide, such as driving cars, using aeroplanes and producing lots of plastic goods. Gases like methane are also being increased through farming practices. This is all contributing to unsustainable rises in global temperature, which is affecting the liveability of the planet. Sea levels are rising (as ice caps are melting), weather patterns are becoming unpredictable and more extreme (with increasing temperatures leading to greater incidences of severe weather, such as droughts and storms) and humans and nature are struggling to adapt.

The challenge for global society is to arrest and reverse this warming trend, principally by reducing the amount of atmospheric carbon, through consuming fewer fossil fuels and looking for ways to remove excess carbon from the atmosphere. This is called Carbon Capture or Sequestration.

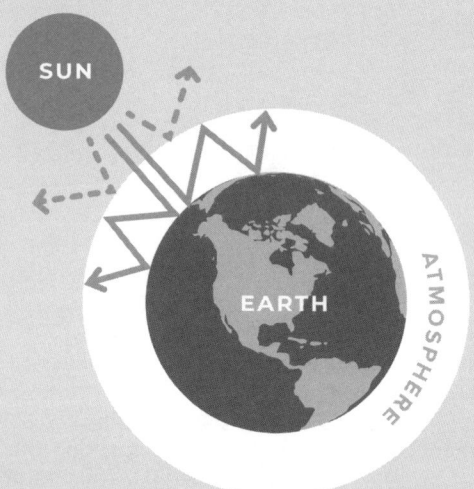

Figure 1.2. The greenhouse effect simplified.

Biodiversity explained in a nutshell

Biodiversity, or *biological diversity*, can refer to both the amount and variety (whether in terms of numbers of individuals and species or total 'biomass') of living 'things' (animals, plants, fungi, and microorganisms like bacteria) present in an area. It relates strongly to the idea of the ecosystems (which also includes non-living or abiotic elements, such as soil, rocks, atmosphere, water, etc.), as part of the complex and interrelated system of living and non-living things that make up the world (or biosphere). The metaphor of a 'web' or 'rich tapestry' is often applied to ecosystems and biodiversity, with greater biodiversity leading to healthier, more resilient ecosystems that are able to bounce back from disturbances. Each individual or species can then be seen as a 'thread' in the living web or tapestry. If they are lost, then the complex system weakens and can ultimately completely unravel. Some species play a particularly important role in shaping the environment or ecosystem and are referred to as 'ecosystem engineers' or 'keystone species'. If they are lost, then the whole ecosystem can quickly unravel and become degraded.

Biodiversity is hugely important to people's health and wellbeing though, of course, individual creatures and species that comprise biodiversity are valuable in their own right, regardless of their value to humans. Indeed, we should see ourselves as *part* of the great family of life (albeit too often a prodigal and selfish family member that does harm to the rest of the family). However, it is also worth reflecting on how our species – humanity – benefits from rich biodiversity and ecosystems. Recently, scientists have come up with the notion of 'Ecosystem Services', which refers to the direct or indirect benefits to human health and wellbeing that are provided by biodiversity and ecosystems. These include: food and materials we get from nature (e.g. wood and wool); clean water and air (including cleaning up of pollution); flood prevention; agricultural land; pollination of our crops and pest control. There are also intangible benefits associated with the positive psychological boosts associated with nature connection (such as reduced stress and physical and mental wellbeing). A key benefit of ecosystems and certain species that photosynthesise (like plants and algae) is their ability to 'clean' carbon from the atmosphere, also known as carbon sequestration, which can be a major solution to climate change. So, rich ecosystems associated with high levels of biodiversity are good for humanity and the biosphere in general. Many attempts are in place around the world to regenerate biodiversity and ecosystem health.

Biodiversity and ecosystems exist in balance with the environment. Sometimes, this balance can be disrupted (such as by natural disasters), leading to a fall in biodiversity and even extinction of some species. This is a natural process, and most of us know about the fate of the dinosaurs and the associated mass extinction that followed environmental change due to the collision of a large asteroid at the end of the Cretaceous period. In fact, there have been five such natural mass extinctions in the geological record. Research suggests that we are now in a sixth mass extinction, with humanity being largely responsible (Kolbert, 2014). Not only are we negatively affecting the existence of other species, but we are unwittingly compromising our own sustainability. Such widely-reported, human-induced threats include:

- **Habitat loss and destruction**: People are building over, encroaching into or taking over land for their own purposes (building, industry, agriculture, etc.).

- **Climate change**: Species have adapted over thousands, if not millions, of years to weather, climate and seasonal patterns. Climate change is shifting these patterns so rapidly that habitats are unalterably changing, and species are unable to adapt quickly enough. For example, hotter temperatures are affecting migration patterns, increasing forest fires, expanding deserts (desertification) and causing coral bleaching in the ocean. Climate change is also associated with ocean acidification, as increases in the amount of carbon dioxide dissolved in the ocean is preventing corals and shellfish from building their bodies.

- **Pollution and waste**: Industries, cities and modern, accumulative lifestyles are creating chemical and plastic pollution, which can poison animals and habitats.

A particular concern highlighted by the Dasgupta Review (2021) is the lack of knowledge that people, especially children, have about the natural world and biodiversity, particularly 'nearby nature' in their home locality. Recent research has found that children are better able to identify advertising signs and fictional characters than local wildlife species (Balmford et al., 2002). This is concerning because if people, especially children, do not come to know nature, they will miss out on the many benefits of 'nature connection' and thus be less likely to look after and regenerate nature. Fortunately, there are signs of a positive shift, with action being taken to remedy this lack of awareness and knowledge. For example, the Save Our Wild Isles campaign (Save Our Wild Isles, n.d.) is a joint campaign by WWF, RSPB and National Trust and is associated with the Wild Isles BBC series by David Attenborough.

The (education) policy context

As we have seen, sustainability, climate change and biodiversity (SCCB) loss represent key concerns facing humanity, yet have not been urgently addressed in UK schools in recent years. The SCCS (Department for Education, 2022) has created a positive space for schools to legitimately take their place as major contributors to a more sustainable future and, in so doing, empowered both teachers and learners *while* improving the quality of learning and wellbeing (both physical and mental).

It is worth putting the SCCS into context, as it has emerged from previous relevant policy initiatives and associated activities over the last few decades. The 2000 National Curriculum (Department for Education and Employment, Qualifications and Curriculum Authority, 1999) had a more significant and overt focus on sustainability, with the government creating the 'Learning Outside the Classroom Manifesto' (Department for Education and Skills, 2006a). The most recent iteration of the National Curriculum for England (Department for Education, 2013) was arguably a step backwards in terms of explicit or overt considerations of SCCB, outdoor learning and learning in the natural environment. Although the aims state that there is time for creative and exciting lessons ranging beyond the National Curriculum; in reality, within a busy curriculum timetable, this has been challenging to achieve. It is important to note that in the other devolved Administrations or Home Nations – Wales, Scotland and Northern Ireland – the curriculum is arguably more conducive to an outdoor learning approach.

However, even where there is no explicit reference to SCCB, outdoor learning and learning in the natural environment in the current Curriculum, there is still plenty of scope to incorporate them through creative interpretation and planning as discussed in Chapters 2, 4 and 5 of this book. Publications such as Bloomsbury's *National Curriculum Outdoors* series (Lambert, Roberts and Waite, 2020), a precursor to this book, also provides ideas and guidance on delivering the National Curriculum programmes of study outdoors. The series includes a book for Key Stage 1 (containing lesson progressions for every curriculum subject in Years 1 and 2) and a book for each year group at Key Stage 2, providing progression in knowledge and skill across the curriculum subjects and Key Stages. Also, organisations such as the National Environmental Education Association (NAEE) have presented 'The Environmental Curriculum: Opportunities for Environmental Education across the National Curriculum for England Early Years Foundation Stage & Primary' (Green, 2015).

By way of further illustration of the possibilities, the Natural Connections Demonstration Project (Waite et al., 2016) was supported by the UK Government (DEFRA and Natural England) between 2012 and 2016. It was led by the University of Plymouth in collaboration with schools and local providers of outdoor and/or environmental education in the southwest of England. The project intended to support schools and teachers to build outdoor learning into their planning and practice *within* the existing constraints (curricular, timetabling, budgetary, etc.). Consequently, the emphasis was placed on learning/activities within school grounds and/or green (or blue) spaces within walking distance of school, with as much as possible to be achieved during normal lesson time, thereby reducing financial burden and disruption to the school day. The project found that far from distracting from the normal business of schooling, using outdoor contexts nearby to teach the National Curriculum actually enhanced learning and pupil engagement. Participating schools saw that limited investments in time, effort and money resulted in massive benefits. This came as little surprise to advocates of environmental and sustainability education and outdoor learning, who are aware of the much broader cause for support discussed below.

Subsequently, the UK Government's 'A Green Future: Our 25 Year Plan to Improve the Environment' (HM Government, 2018) and 'Landscape Review' (Department for Environment Food and Rural Affairs, 2019) both identified a need to increase and enhance opportunities for children and young people to engage with nature. Drawing on the success of the Natural Connections Project (see Chapter 3), these policies provided the impetus for the 'Nature Friendly Schools' project, 2019-2021. Funded by the UK Government, including the Department for Education, and led by The Wildlife Trusts in partnership with a range of other organisations, the goal was to demonstrate school-led outdoor learning, particularly in disadvantaged areas. Then, in 2021, the UK Government hosted the 26th Climate Change COP in Glasgow and launched the SCCS, which builds directly on these recent positive developments.

Whilst the curriculum is not impacted directly, the SCCS itself takes things further and has several dimensions. One dimension is the promotion of Sustainability Leadership in each school with the aim of producing a Climate Action Plan, leading to Climate Action Awards for schools. A second aspect is the National Education Nature Park concept, with the totality of schools' grounds across England forming a kind of disaggregated 'National Park' to engage young people with the natural world and directly involve them in increasing biodiversity. Thus, each school can start to think of their school grounds as a parcel in this wider 'National Park' provision, and start to use it accordingly (as a natural space available for teaching and learning, as well as a resource for leisure, and promoting physical and mental health and wellbeing).

Broadening horizons: Justification beyond the curriculum and education policy

Whilst the SCCS is an exceedingly encouraging and welcome development, the potential of this policy shift to enact the kinds of change required has been scrutinised, questioned and found somewhat wanting. Dunlop and Rushton (2022) produced a critical examination of the UK Government's strategy which they feel provides a depoliticised and 'placebo policy', which is seen to fall short of the radical shift required by activists and educators, including schoolteachers. Similarly, Greer and Glackin (2021, p. 18) identify a current 'climate change education policy gap'. Likewise, Teach the Future, the youth-led pressure group advocating for education to address the Climate Emergency, provided an extensive critique of the strategy (Teach the Future, 2021) and have gone as far as to present a 'Climate Education Bill' to Parliament demanding greater and more radical change to the curriculum, management of the school estates, and teacher education, in response to the Climate Emergency.

So, why is it that such educators are calling for a more critical or appropriate educational response to the needs of learners, wider society, nature or the planet? What kinds of wider educational outcomes are they calling for? There is a broad consensus emerging across the world as to what is required. For example, the 'Berlin Declaration on Education for Sustainable Development' (UNESCO, 2021) emerged from the World Conference on Education for Sustainable Development convened by UNESCO in 2021. This Declaration identified education as key to the 'urgent action [that] is needed to address the dramatic interrelated challenges the world is facing, in particular, the climate crisis, mass loss of biodiversity, pollution, pandemic diseases, extreme poverty and inequalities, violent conflicts, and other environmental, social and economic crises that endanger life on our planet.' (UNESCO, 2021, p. 1). Alongside the Declaration, UNESCO has created a Roadmap for achieving Education for Sustainable Development (UNESCO, 2020).

The sustainability agenda is very large and complex, integrating a range of inextricably interrelated issues (hence the complexity of there being 17 SDGs). Our book is intended to be a contribution towards achieving this broad vision of a sustainable future, with focused attention on the specific issues of climate change and biodiversity through Outdoor Learning, which are themselves interrelated and synergistic. This allows us to consider the particular contribution of, and benefits to, education and schools and their grounds in relation to SCCB.

A good starting point would be to broaden consideration of learning outcomes beyond mere 'test success', towards a more holistic vision. Rickinson et al. (2004) provided a very useful review of research on the benefits or impacts of learning outdoors, particularly in relation to formal learning. They frame their analysis according to:

- cognitive (intellectual domain)
- affective (domain of emotions, attitudes and values)
- social/interpersonal (or personal and social development)
- physical/behavioural impacts.

The Natural Connections Project (see Chapter 3) considered, and found significant evidence for, the learning benefits across all of these domains. The key message is that addressing the other important (but often ignored) dimensions of learning, alongside the cognitive/intellectual, better reflects the needs of individuals and society. In so doing, you can also enhance test scores because learners are more engaged and have understood the subject matter more deeply through real-world experiential learning.

Similarly, the UK-based charity Learning Through Landscapes lists five key benefits of outdoor learning (Harvey, 2022):

- children's mental health and wellbeing improves
- children's relationship with nature improves
- education is more inclusive
- curriculum learning is enhanced
- child development is enhanced.

The Children & Nature Network, a new global movement, advocates such a holistic perspective, extending it to consider not only the needs of the individual and society but the natural world too. In particular, their Greening School Grounds and Outdoor Learning Project (in partnership with other organisations) identifies five particular benefits or outcomes which neatly encapsulate the key messages of this book (see Figure 1.4). As their website states, 'nature-filled school grounds enhance children's play, health and wellbeing – and help improve climate resilience, community engagement and educational outcomes' (Children & Nature Network, n.d.) – a win-win-win scenario.

Figure 1.4: The dimensions or benefits of Greening School Grounds & Outdoor Learning, reprinted with permission from The Children & Nature Network.

As we try to stress throughout this book, such benefits are interrelated and mutually reinforcing – hence *holistic*. However, it is worth trying to tease out some of the arguments that feed into this broad vision. The case for teaching SCCB from a range of perspectives, especially outdoors, can boil down to the following arguments or grounds:

1. Cognitive or intellectual

Taking intellectual outcomes first (as this is perhaps the key message that will persuade the powers that be in the prevailing, but somewhat constraining, context), the argument is that teaching about real-world issues (such as climate change and biodiversity) in real-world contexts (such as outdoors and in nature) can significantly enhance and enrich the educational experience and achievement of learners. Typically, education tends to foreground the 'cognitive' – the intellectual process of learning 'content (or declarative) knowledge' – knowing *what* (i.e. subject-specific facts that can be e.g. reproduced in a test); and/or 'procedural knowledge' – knowing *how* (i.e. procedures, skills and/or competencies that might be generic, such as critical thinking, or subject specific). Of particular relevance in relation to the latter is 'enquiry-based learning' or 'learning through enquiry'. This is often associated with science (as 'Working Scientifically') and geography (as 'geographical enquiry') but has much broader cross-curricular relevance.

Similarly, broader or more generic 'Thinking Skills' emphasise the educational value of developing intellectual competencies in critical and creative thinking and problem solving. These can promote exam success and job performance but, more importantly, are precisely the skills and competencies that are demanded of our citizens in the twenty-first century. For example, creativity 'involves the capacity to generate, to reason and to critically evaluate novel ideas and/or imaginary scenarios. As such, it encompasses thinking through and solving problems, making connections, inventing and reinventing, and flexing one's imaginative muscles in all aspects of learning and life' (Cremin, 2017, p. x). Cutting and Kelly (2014) have provided a great overview of opportunities to teach primary science creatively. Three chapters in particular would be very relevant: Chapter 10: Sustainability and Primary Science, Chapter 11: Teaching Science Outdoors and Chapter 13: Teaching Controversial Issues in Science.

2. Environmental citizenship, participation and stewardship

The many interconnected issues that are implicated in sustainability, including the climate and biodiversity crises, demand informed and committed citizens participating collaboratively to bring about societal change and to act as stewards of the environment. The National Curriculum expects schools to prepare learners morally and as citizens, and it is hard to think of more pressing moral issues than those implicated by SCCB. In addition, while not often emphasised, the current National Curriculum has the requirement that:

> 'Every state-funded school must offer a curriculum which is balanced and broadly based and which:
>
> • promotes the spiritual, moral, cultural, mental and physical development of pupils at the school and of society, and
>
> • prepares pupils at the school for the opportunities, responsibilities and experiences of later life.'
>
> (Department for Education, 2013)

Arguably this underpinning rationale could – indeed *should* – be interpreted to integrate considerations that relate to both SCCB (as key issues that are affecting, and will affect, learners as citizens throughout their lives) and outdoor learning (that particularly supports spiritual, moral, cultural and physical development). These broader personal and societal justifications are developed further below.

However, unlike the phrasing in the National Curriculum document which sees this as preparing them for 'later life', we would stress that we should be preparing them as active citizens in their own right, with agency and ideas, right now. We only have to see the incredible passion of young people, such as Greta Thunberg and those involved in the Teach the Future coalition, to know that many young people are *already* active citizens in their own right. As Dewey (1998) suggested at the turn of the last century, if you want learners to learn about democracy, you have to give them opportunities to learn through participating meaningfully in democratic processes. Therefore, allowing learners to meaningfully engage in positive activities to enhance biodiversity and climate action on their own school grounds represents a powerful argument for integrating SCCB in outdoor learning in our schools.

It is worth asking what the barriers might be that hinder personal involvement with SCCB. Tasquier, Pongiglione and Levrini (2014, p. 822), talking principally about climate change (but in ways that are applicable to biodiversity and sustainability too), identified three main barriers:

a) *Not individual but collective:* The individual does not see how their individual actions can have any impact on issues which are clearly demanding collective solutions, or those by big institutions like the state.

b) *Too big or too small:* Either the issue is perceived to be too overwhelming for the individual to have any meaningful impact, in which case ignoring the problem is a defence mechanism to avoid feelings of fear, guilt or hopelessness; or the issue is not at the forefront of their attention as they are not facing immediate implications in their everyday life, meaning they can ignore it.

c) *Too far:* The issue is felt to be too far removed from them (in space i.e. affecting people 'over there' but not here); and/or in time (i.e. will not have immediate impact, meaning the dangers lie in the future and can therefore be ignored).

Having the opportunity to engage in a meaningful way as an individual and member of a local school community through SCCB in outdoor learning on the school grounds could go a significant way to overcoming these barriers and nurturing environmental stewards.

But we also have to be mindful that SCCB are controversial issues, 'Wicked Problems' (Rittel and Webber, 1973), and/or complex socio-scientific issues. That is to say, it is not necessarily obvious, nor will there be immediate consensus, as to the nature of the problem, or the requisite solutions. Society is made up of many people with different values and attitudes, and part of the demand of participatory citizenship is an ability to cope with different perspectives and deal with conflict in appropriate ways. So, we also need to equip our learners with the knowledge and skills to be able to critically interrogate different perspectives and to empower them to arrive at their personally authentic response to these issues, and so become informed citizens and capable of respectfully arguing their position.

Again, these important life lessons are best learned through real-world, experiential engagement in collaborative problem solving. To do so will contribute to learners' intellectual development (see above) and, crucially, contribute to their personal and social development and empower them as agents of change. This will enhance their self-efficacy and self-esteem, key dimensions of positive psychology. The progressions included in Chapters 4 and 5 of this book are designed to address these points.

3. Positive psychology, wellbeing and spiritual, social and emotional development

The final cluster of justifications presented here revolve around questions of what it means to be human, and how we might best support learners to flourish. Here, the emerging field of 'positive psychology' is helpful. Feelings of agency, self-efficacy, meaningful engagement and collaborative effort all contribute to positive, affective states and feelings of fulfilment, wellbeing and self-esteem, as well as promoting personal and social development, citizenship and stewardship. Recently, psychology has also revealed the importance of connection to nature, which appears to be a natural and universal need of humans, called the biophilia hypothesis (Wilson, 1984). A similar case has been made for the importance of a 'sense of place' for healthy identity formation and functioning (Altman, Low and Chawla, 1992, pp.63–86). Being experientially immersed outdoors in nature in familiar places such as school grounds could contribute positively to the development of positive 'place attachment' and nature connection.

Figure 1.5: The Love Living Goals © Warwick, Warwick and Nash 2017, reprinted with kind permission from Paul Warwick.

However, we also need to be mindful of the danger of negative emotions or effect. As noted above, feelings of fear, guilt, powerlessness and/or hopelessness become barriers to participation and result in negativity and stress. A key issue is the mental health implications involved in addressing often alarming issues that can seem overwhelmingly large and beyond the ability of individuals to address. This can lead to anxiety, which arrests, rather than promotes, positive engagement and problem solving. The concept of 'eco-anxiety' speaks to the real dangers of frightening people, learners included, with the enormity of global problems, such as SCCB, that they feel they have no capacity to address. Indeed, some have argued that it is best *not* to seek to address these issues with younger children. However, we think that it is more productive to engage with these issues at a scale that is appropriate i.e. the local, small scale – such as the school grounds – where learners can contribute to, and witness, positive change. This relates to the notion of 'place-based education'. Consequently, it is important to emphasise an education of hope. Hicks (2014) has specifically written about the need for such a perspective in the time of climate change and the transition to a post-carbon world. It is also crucial to make it accessible – as attempted by 'Never Too Small' and their associated 'Love Living Goals', which represent a child-friendly version of the SDGs (Warwick, Warwick and Nash, 2017).

To conclude this introductory chapter, we would like to share the metaphoric 'Bicycle Model' that was devised through an extensive literature review on climate change education (Cantell et al., 2019), which we feel works well for all aspects of SCCB. The model shows how, as educators, we need to think about multiple dimensions of learning (knowledge, thinking skills, motivation, identity, attitudes and values, participation and collaboration, etc.) in a manner that sees them working together to achieve the goals of a hopeful, future-oriented and environmentally-friendly solution to the SCCB challenges of the present and future. In the progressions included in this book, we have tried to take account of all these dimensions and it is well worth bearing this model in mind as you engage with the rest of this book.

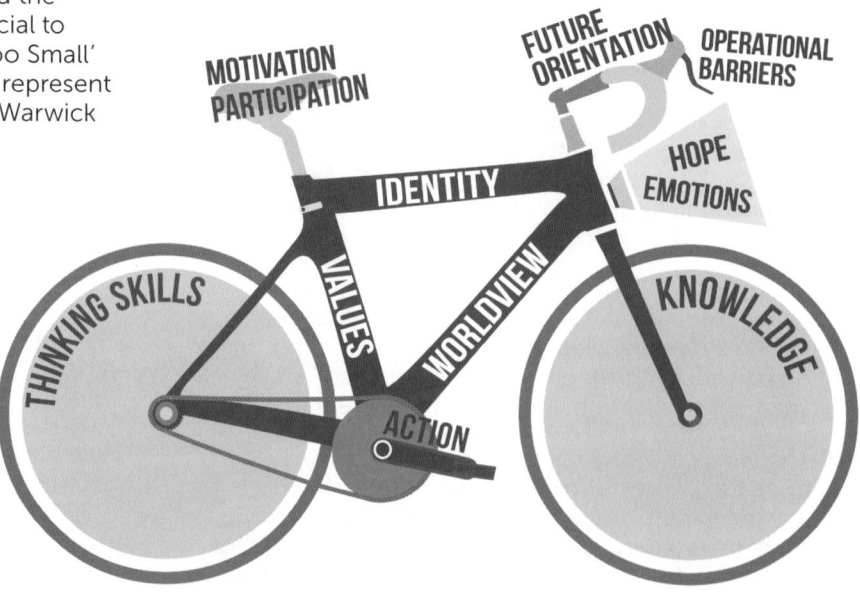

Figure 1.6: The Bicycle Model (Cantell et al., 2019, p. 719), reprinted with kind permission from Hannele Cantell.

Chapter 2
The role of leadership in affecting sustainable change

Introduction

The Sustainability and Climate Change Strategy (SCCS) proposes to 'increase opportunities for all children and young people to spend time in nature and learn more about it, as well as becoming actively involved in the improvement of their local environment' (Department for Education, 2022 p. 8). With the requirement to deliver a whole-school approach to promote greater understanding of the importance of nature, sustainability and climate change, and translate this into positive actions and solutions, the strategy acknowledges that coordination and leadership will have the greatest impact on success.

During the last decade, there has been a significant increase in the number of schools developing and implementing outdoor learning as part of their curricular and extra-curricular offers, with a variety of different models and approaches adopted as part of this process. However, in some cases, the longevity or sustainability of these approaches has been a challenge. This chapter explores the role of leadership in developing a whole-school approach to outdoor learning using a sustainable leadership model that has been proven to be effective in schools.

Embedding outdoor learning in your school is an ongoing, evolving journey, underpinned by whole-school culture, policy and plans. The 2022 SCCS provides a unique opportunity for schools to embed the principles of climate change education through outdoor learning curriculum models, as well as developing commitment to a wider sustainable approach.

A positive, knowledgeable and confident subject leader can inspire the whole-school team to overcome barriers to adopt an outdoor learning approach. They can provide clear expectations and support to achieve this, through a clear programme of consultation and continuing personal and professional development, training, evaluation and review.

A successful model, key to many productive approaches, has been to implement a leadership process that not only embraces a sustainability and climate change curriculum but also ensures an effective and manageable approach to delivery. A commitment to leadership is at the centre of this approach.

Figure 2.1: Wild Tribe Outdoor Learning: Sustainable Leadership Model (Roberts, 2022).

This model (Figure 2.1) illustrates a continuous, cyclical process, which a school might adopt to implement a strategic plan for Outdoor Learning. This model underpins the development of subject leadership within a primary setting through Outdoor Learning. It has been developed and successfully trialled by the Arena Wild Tribe Leadership Training Programme and has proven to promote a sustainable approach to whole-school outdoor learning:

1. It starts by **establishing the ethos, mindset and vision** for sustainable outdoor learning. This might, for example, include a sustainability charter or intent statement, linking to the SCCS and wider curriculum delivery.

2. It continues **developing or adopting a whole-school approach**, including an audit of existing provision as a baseline, from which whole-school confidence and practice can be built.

3. The development of the school grounds is explored in more detail in Chapter 3 of this book and supports the next step in the cycle, which is to **create a 3-year sustainable site plan**. The establishment of your school grounds will be based on the actions and opportunities identified in the baseline audit. It might include the development of the school site to provide curriculum-based learning opportunities. Arena suggests utilising Sobel's seven principles (2008) as a guide and inspiration.

4. The next stage is to **upskill the staff and stakeholders by providing targeted CPD and training for the staff**, to improve knowledge, confidence and skill progression across the Key Stages. This will be coordinated by the outdoor lead practitioner and supports the whole-school ethos and school development plan.

5. Next is a **consideration of the role of the pupils** and how they will contribute to the outdoor learning provision. This includes the site provision, perhaps through pupil leadership, school council or climate change ambassadors.

6. The **development of a sustainability and climate change curriculum** across the Key Stages and for each year group is the next step. This might also be influenced by the pupils. Examples of the types of progressions that could be embedded within a primary setting can be found in Chapters 4 and 5 of this book, where we have detailed learning progressions to illustrate how climate change and sustainability education can be implemented at Key Stage 2. The aim is to develop a year-round curriculum, engaging with nature and developing nature connectedness and understanding of the impact of seasonal and climate change, alongside teaching the key concepts of climate change and sustainability.

7. The final stage is to **evaluate the impact of the plan**. Although evaluation will be an ongoing process as part of a three-year plan in line with current school leadership practices, this step offers the chance to reflect on the impact of the whole cycle.

Each part of the model plays a key role; if one or more parts of the cycle are not fully implemented, outdoor learning becomes fragmented or even ceases to exist.

Other providers who support schools in the development of a whole-school approach to outdoor learning include:

- Learning through Landscapes

- Eco-schools

- Transforming Schools

- World Wildlife Fund

- Natural History Museum.

Each of these organisations have comprehensive guides for schools to support the development of a whole-school approach to outdoor learning.

1 Exploring the sustainable leadership model: Ethos, mindset and vision

This section explains the importance of developing a knowledge-based, whole-school ethos, mindset, vision and approach to outdoor learning and sustainability, climate change and biodiversity. It highlights the impact this can have within schools on both children and teachers.

Developing an ethos for your school is of key importance and underpins the delivery of a long-term commitment to outdoor learning. To implement change, schools need to start to make it part of the culture of the school, build the mindset with the pupils and staff, and ensure it becomes part of the ethos.

In response to the Cornwall Schools' Eco Conference in June 2021, every school in Cornwall has signed up to a Green Charter for schools. This charter 'commits all Cornwall's schools to working together for a greener, cleaner, fairer future. Being greener together means being part of a bigger community. Across Cornwall's schools we are educating children and young people about climate change, empowering them to take action' (Cornwall Green Schools, n.d.).

The introduction of the SCCS has seen many schools develop a sustainability charter to ensure that their commitment to climate change is evidenced and becomes part of the school and/or Multi-Academy Trust.

An example of this is Coads Green Primary School, Cornwall, which has a sustainability charter:

To embed a whole-school approach to outdoor learning, all stakeholders must understand the impact that it can have and the rationale behind such an approach, as part of the cyclical development process. So let's consider, why take learning outdoors?

Reflection

Close your eyes...

Think back to an early childhood memory.

Was it indoors or outdoors?

What made it memorable?

Why do you think this is?

Many of our memories will have been made while outdoors. As covered in Chapter 1, the Department for Education and Skills published the 'Learning Outside the Classroom Manifesto' for England in 2006. This was and is a shared vision to raise achievement through an organised and powerful approach to learning, in which direct experience and experiential practice are of prime importance. It reinforces what you may already have found when answering the reflection questions – outdoor experiences are often the most memorable. More recently, the Department for Education has reinforced the importance and impact of regular outdoor experiences to not only build nature connection but to promote children's physical and mental health (Department for Education, 2022).

Outdoor learning provides multi-sensory experiences, elements of risk and challenge and real-world 'using and applying' experiences often called *authentic* learning. What we do, see, hear, taste, touch and smell gives pathways to deeper learning, the potential for which is maximised by adopting a powerful combination of physical, visual and naturalistic ways of learning (Department for Education and Skills, 2006a). This multi-sensory, experiential approach provides the opportunity for teachers and practitioners to offer relevant tasks that have a real purpose. It allows them to introduce, reinforce or make connections with wider learning. This may influence and impact future values, behaviour, beliefs, lifestyle and even work, by linking learning with feelings (Department for Education and Skills, 2006a; Natural England, 2022; Dasgupta, 2021; Altman, Chawla and Low, 1992).

Faced with a generation of children perceived as 'indoor children', schools face significant challenges to ensure children develop emotional resilience. The benefits of being outdoors are now widely recognised; positively impacting confidence, social skills, language and communication, motivation, concentration, physical skills, knowledge and understanding, as well as improving engagement and educational outcomes within the school setting (Waite, 2020; Marchant et al., 2019; Mann et al., 2022). The evidence to support the link between outdoor learning, health and wellbeing and pupil attainment is robust. It is important to note that the culture, ethos and environment of a school also impact the health and wellbeing of staff (Public Health England, 2014, Waite et al., 2016).

Consulting all stakeholders (including the children and parents) is important. Ensuring all stakeholders have a shared, knowledge-based vision is key. If quality outdoor learning is to be delivered regularly by all teachers, then embedding outdoor learning across the whole school must be part of the culture and ethos of the school (Bosevska and Kreiwaldt, 2019). This can be done by regularly reinforcing the vision and its values, perhaps through a half-termly focus, to ensure the vision becomes part of the culture and everyday language of the school.

2 Whole-school approach to sustainability and climate change awareness

This section explains how to develop a whole-school approach to outdoor learning and SCCB, what it might look like, its importance and the impact it can have within schools.

Creating a common understanding of the impact and rationale of an outdoor learning approach as a medium for climate change education can be shown using an 'intent statement'.

Generally, a clear curriculum intent will provide evidence of a coherent and well-sequenced curriculum for the 'quality of education' being offered. This can prove particularly useful during an Ofsted inspection 'deep dive', where inspectors talk to senior leaders to find out whether the curriculum being offered is 'broad and balanced' or for Academy schools (which do not have a statutory obligation to follow the National Curriculum), that what is being taught is at least as ambitious as the National Curriculum.

Reflection

Consider this sentence from the Education Inspection Framework:

"

'The school's curriculum is rooted in the solid consensus of the school's leaders about the knowledge and skills that pupils need in order to take advantage of opportunities, responsibilities and experiences of later life.' *(Ofsted, 2019)*

"

How embedded is your outdoor provision within the whole-school curriculum?

Reflect on the intent and purpose of your outdoor curriculum in light of the SCCS guidance. Doing this enables you to make informed choices about what experiences to include, how they build over time across the curriculum subjects and how to identify broader learning experiences, such as off-site experiences or community involvement.

Start by establishing your outdoor learning and SCCB curriculum principles, which will reflect your school's ethos, values, context, pedagogy and needs. Explain what pupils will experience as they move through school sequentially. Make progression of knowledge and skills clear, linked to the school's long-term planning, and outline how it supports a broad and balanced curriculum by stating what the outdoor learning curriculum covers, how it covers it and in how much depth.

Reflection

To consider in the context of your own setting:

- Why have you adopted an outdoor learning approach? E.g. to support mental health and wellbeing, to support a school development plan, etc.

- Is the outdoor learning experience providing pupils with the building blocks of what they need to know or do to succeed?

- How does the outdoor curriculum intent relate to your setting's ethos?

- How strong is your outdoor learning curriculum delivery and what makes it that way?

- Make a mind map of the knowledge and skills that you feel pupils need to take advantage of opportunities and responsibilities as they get older. How can these be evidenced outdoors?

Such reflection raises these often-mentioned questions:

- Is outdoor learning a separate subject or an approach to delivering curriculum subjects?

- Might having a separate intent statement for outdoor learning segregate the outdoors from the totality of the learning experiences (Education Scotland, 2020)?

A statement inserted into other policies, such as teaching and learning policies, about the value and need for outdoor learning will reinforce to the whole school the commitment to an outdoor learning approach.

With a whole-school approach, all stakeholders work together to create a school-wide model of outdoor learning to support a greater awareness of the challenges that SCCB present.

Audit of provision

How to create a baseline audit of the setting, including resources, staff levels of confidence and setting provision

Conducting an audit provides information about where you are with regard to outdoor learning and provision at your school. This audit should be regularly reviewed, noting any changes or actions for your outdoor learning development plan, as part of the wider school improvement plan.

In summary:

- Carry out a detailed audit of all areas of provision including staff, community, external providers, school site and resources.

- Establish the long-term goals for outdoor learning, considering what you aim to achieve.

- Put an action plan in place to cover all areas of a whole-school approach, key stakeholders, teaching and learning, leadership and vision.

- Continually research outdoor learning approaches in different contexts.

- Visit other school sites and network with other subject leaders.

3 Creating a 3-year sustainable site plan

To develop a 3-year site plan, the school needs to consider its key components

Once you have completed an audit of your site and your existing provision, the next step is to develop a 3-year sustainable site plan. Before developing this plan, consider: What is an effective outdoor learning environment? Why do we need one?

David Sobel, a place-based educator, identified a set of seven play themes that emerge when children experience free time in nature. These themes are adventure, fantasy and imagination, animal allies, maps and plans, special places, small worlds and hunting and gathering (Sobel, 2008). When applied in an outdoor area, these themes make learning more effective. There is no doubt that a natural space leads to richer imaginative play, increased physical activity, calmer, more focused play and positive social interactions, but what if you only have a concrete area or a field with no trees?

Your site plan needs to meet the needs of your site and the curriculum you wish to provide for your children. The case study below explains how one school developed a site plan using the principles of Sobel's model. To ensure that it was embedded within the whole school, they linked the plan to the key curriculum drivers and their intent statement.

Case study

Sithney Community Primary School, Cornwall (further information on this plan is available in Appendix 3)

In developing a 3-year plan, Sithney School started with the school map and zoned the areas into key themes linked to the curriculum intent and school values. They identified four zones:

- Active Zone
- Discovery Zone
- Earth Lab
- Inspiration Zone.

Once these had been identified, the school looked at what was in these zones, what the ecological impact of change in these areas would be and what action needed to be taken to develop the zones further. For example, in the Active Zone, there were newly planted trees, wildflowers, a sports field and an active playground. The learning focus for this area was for children to be able to demonstrate the key values of perseverance, kindness, resilience and respect.

To ensure this could take place, the following actions were identified:

- Promote use of the centre of the field and playground for sport and athletics.

- Provide maintenance to newly planted trees, including clearing the area around the base (animal allies).

- Complete a wildflower survey and promote wildflower growth on hedgerows and periphery of field (animal allies).

- Allow natural growth and die-back of fungi in the centre of the field during autumn.

The ecological impact of these actions was to promote the biodiversity in this zone. Taking this approach, the same theory was applied to all the zones, which developed a clear plan.

The school next decided on a timescale and agreed on what was achievable in the time allocated to ensure the plan could be successfully implemented.

Matthew Birchall, Climate Literacy Lead Teacher and Coordinator of PE and Outdoor Learning, Sithney Community Primary School

4 Continuing personal and professional development

Whilst there is no requirement to hold a qualification in outdoor learning, some schools choose to upskill their teachers through staff meetings, or with focused continuing professional development (CPD), continuing personal and professional development (CPPD) and in-service training (INSET) days. This can provide teachers with the knowledge and skills to confidently offer engaging, effective and enjoyable outdoor learning.

The advantages gained from upskilling in these ways differ. With CPD or CPPD, perhaps only one or two members of staff who have a professional development interest or need, or a personal curiosity or enthusiasm, attend specific training. Such training may occur over one day or a series of days, perhaps leading to an accredited qualification. For example, a member of staff with an interest in outdoor learning, or with this area of responsibility, may attend a climate change awareness training day. CPD training days often occur at a geographically central venue, with individual participants attending from a variety of settings, building relationships with others with similar interests and professional needs, resulting in great networking opportunities. Good CPD deepens the professional understanding of the participants. For example, in the example of a climate change awareness day, upskilling the participants can affirm the benefits of an outdoor learning approach. Research shows, however, that not only does it take time for CPD to embed pedagogical changes, but the time required to undertake such training is a challenge for schools, particularly within a busy timetable and with existing teacher workloads (Adey et al., cited in Marchant et al., 2019, p. 18). The

sharing of knowledge and content from CPD training to colleagues back at the setting (not all of whom will be 'on board' with a pedagogical change or share the same interest and drive), relies on providing sufficient time and opportunities to do so, and may not provide sufficient 'immersion' to make an effective difference or impact.

An alternative is taking a whole-school approach to training with an INSET day or twilight staff meeting. The purpose of an INSET day is to present CPD training that is deemed important for the whole-school staff. It's an opportunity for inspirational training, with the whole staff working together with shared knowledge and approaches, to debate and agree on strategies to support whole-school improvement, leading to better outcomes for pupils.

- Planning for continuity, progression and assessment.

- Developing advocacy skills in primary outdoor learning.

These modules support teachers in developing their pedagogy in outdoor learning and whole-school approach to the subject. As a result of this course, teachers have reported they feel more confident in leading outdoor learning across the school and feel empowered to bring about the changes required to implement a development plan to support outdoor learning within their setting.

Arena regularly delivers seminars and conferences on outdoor learning to ensure that teachers remain up to date with the latest developments in line with the following Education Endowment Foundation recommendation: 'effective professional development includes both initial training as well as high-quality follow-on support' (2021).

Michelle Roberts, Wild Tribe

5 The power of children's voices in taking action for change

This section explores the rationale for pupil voice when constructing a curriculum and whole-school approach, to enable empowerment and taking action.

Article 12 in the United Nations Convention on the Rights of the Child states that 'every child has the right to express their views, feelings and wishes in all matters affecting them, and to have their views considered and taken seriously' (UNICEF, 1992). Research shows that young people can feel more optimistic about climate change when they know there is action that can be taken (Ojala, 2016; Li and Munroe, 2019). Pupil voice offers a valuable perspective on the experiences of not only what it's like to be a pupil in school but also what it is like to be a child at this time. Children are confronted with social narratives of crisis and threats relating to ocean plastics, vanishing habitats, reduced biodiversity, species extinction and extreme weather. Gaining insight into the lived and shared experiences, views and understanding of young people provides the opportunity to turn anxiety, fear, frustration and inaction into efficacy, participation, agency and empowerment through action for change (Gutiérrez, 1994).

Case study

St Alban's CE (Aided) Primary School, Hampshire

"

'Nature gives and gives and we can never measure up, but we must do what we can. We want to get more people helping nature so there is a chance for nature to have a chance. We are fascinated by the natural world and there is so much to uncover – new species are being discovered every year. We want other people to understand and experience how nature makes us feel.'

"

The Tree Council's Young Tree Champions Junior Ambassadors 2023 (Frankie, Immy, Maisy and Matthew)

Schools with a strong commitment to pupil voice provide pupils with meaningful opportunities to express their views as valued members of the school community. Positive actions as a result of pupil voice can be made visible on the school websites, classroom noticeboards, whole-school action planning, school statements and newsletters.

Case study

Droxford Junior School, Hampshire (the full version of this case study is included in Appendix 2)

The school website emphasises the importance of children being fairly represented and consulted on what happens in their school life (e.g. the learning curriculum), via the school council. This has key actions that form their focus for the year. These key actions include, 'ensuring children are an integral part of the school grounds development' and 'helping to protect the environment' (Droxford Junior School, n.d.). There are also key tasks for the school council, including 'being more eco-friendly and more sustainable' and 'develop[ing] the way in which we learn outdoors' (Droxford Junior School, n.d.). In taking action for change, the children have formed a 'Wilder Committee' in association with their local Wildlife Trust, to improve habits on the school grounds. The children successfully applied for a funding grant, and have built bug hotels, bee boxes, insect houses and bird boxes, as well as engaged the wider school community through Dig Days which have enabled the school, through a huge amount of parent power, to establish much of the outdoor planting and wildlife areas.

The modelling or showcasing of 'taking-action for change' and 'making environmentally friendly choices' influences others to adopt similar choices (Klockner, 2013). Seen as a 'shared good practice', such modelling may help to mitigate the risk of individuals feeling disempowered by the belief that they are making only 'little differences' or taking 'small steps', given the enormity of the climate crisis. The term 'individual' in the expression 'individual action' should never be underestimated, since what is collective action without individuals making individual differences to greater effect? As Greta Thunberg states, 'No one is too small to make a difference' (Thunberg, 2019). Examples of pupil action might include:

- Collaborative action to raise awareness of local issues, such as biodiversity lack or loss, then taking action to change that. This could be by rewilding, tree planting or planting pollinator gardens in school or in nearby open spaces and community gardens.

- Taking action to reduce waste and litter by saving water, reducing plastic use and promoting recycling and composting, as well as contacting local representatives about policies.

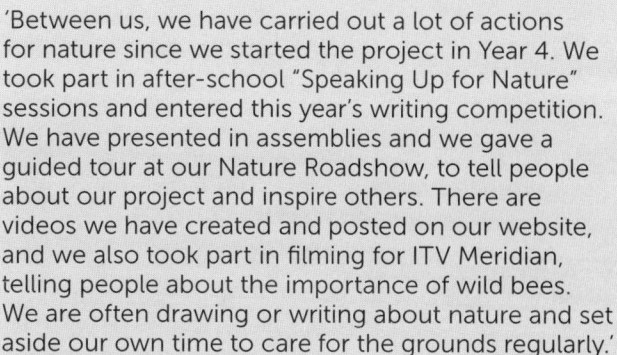

Case Study

St Alban's CE (Aided) Primary School, Hampshire

❝

'Between us, we have carried out a lot of actions for nature since we started the project in Year 4. We took part in after-school "Speaking Up for Nature" sessions and entered this year's writing competition. We have presented in assemblies and we gave a guided tour at our Nature Roadshow, to tell people about our project and inspire others. There are videos we have created and posted on our website, and we also took part in filming for ITV Meridian, telling people about the importance of wild bees. We are often drawing or writing about nature and set aside our own time to care for the grounds regularly.'

'In the past, we have shared our ideas through newsletters, our school website, assemblies and the Club Space. We've made videos, written poetry and campaigned outside of school. As well as carrying on with these things, we'd like to add an outside display board that showcases what we are doing. We need to "go public". We liked appearing on ITV Meridian and saw that it can have a big impact, so we think we should make more use of TV. Writing social media posts is on our list because we see that many people use it and it can spread the message to a wider community of people. Lots of children across the school have been involved in the project but we'd now like each house to champion a mini project involving biodiversity or reducing pressure on the environment. We want to help Key Stage 1 lead more things and help them achieve Hedgehog Friendly School status. [This is an award from the British Hedgehog Preservation Society for taking positive action to provide a nurturing habitat for hedgehogs.] However, this year we would like to bring the adults in. We want to share our thoughts and feelings with them, so they understand the magic of nature.
We think we should contact councillors and join up more often with local organisations. We'd also like to bring our parents into school, to help us make changes to our grounds so we can leave a legacy but also to inspire them to make changes at home.'

❞

The Tree Council's Junior Tree Champion Ambassadors 2023 (Frankie, Immy, Maisy and Matthew)

Although every setting is unique there are a range of ways pupil's views, feelings and wishes can be gathered to encourage children to be agents for change, reflecting the importance and value of pupil voice and the commitment to listening to the voices of the children in the ethos and values of your school.

Young people are aware that we are in the midst of a climate change and biodiversity crisis. So how might pupil voice contribute to and shape a curriculum that raises awareness of climate change and biodiversity and balances the delivery of essential knowledge, without creating a culture of 'climate anxiety' and hopelessness, leading to inaction? A curriculum based on a 'pedagogy of hope' may empower children, enabling them to make a difference in the knowledge that they can be agents of change.

6 Developing a sustainability and climate change curriculum

This section reiterates the importance of experiential approaches and explores the potential within the National Curriculum for SCCB education.

In the Early Years Foundation Stage and at Key Stage 1, children explore the natural world, local environments and microhabitats of plants, animals and living things. Such early nature experiences provide a baseline from which positive behaviours towards nature and compassionate connection can develop (Kidd, 2020). This connection may well last into adulthood and promote future environmental stewardship (Broom, 2017). Providing real-life investigations and hands-on activities as a focus for learning helps shape children's attitudes towards nature. This enables them to understand the importance and role of nature in sustainability and the impact of climate change. By adopting a hands-on approach outdoors through direct experiences of nature, young people can be empowered to not only understand 'the facts' but also to develop a greater appreciation of nature. This is achieved through experiential, interactive discussions about environmental problems and solutions, offering opportunities for critical thinking and communication skills, creative thinking and collaboration, showing children that they can be agents of change (Malone & Waite, 2016).

Case study

Arena Wild Tribe Earth Tribe

Arena Wild Tribe have developed a SCCB programme, Earth Tribe, to be delivered for schools outdoors. The core themes running through the scheme include developing citizen scientists, climate change action, sustainable education and taking action at school. There is also an Earth Tribe Leadership Award for children in the school linking school action to the National Education Nature Park.

The *Sustainability and Climate Change Outdoors: Key Stage 2* book supports this programme.

More can be found out about this award by visiting: www.arena-schools.co.uk/wild-tribe.

The Department for Education aims state that the National Curriculum is just an element in a wider school curriculum, aiming to promote 'the development of pupils' knowledge, understanding and skills' (Department for Education, 2013b). Thus, the opportunity arises to range beyond the National Curriculum programme of study specifications at Key Stage 2, to embed climate change education through a variety of means, increasing opportunities for children and young people to learn from and connect with nature and improve biodiversity in their schools.

But where might SCCB appear in the National Curriculum for England at Key Stage 2?

At lower Key Stage 2, prior knowledge of natural processes is developed to include working scientifically to 'talk about, test and develop ideas about everyday phenomena and the relationships between living things and familiar environments' (Department for Education, 2013b, p. 148). This is alongside learning about function, relationships and interactions, as well as observing changes over time and noticing patterns. The guidance recommends that the local environment be used to provide experiential scientific enquiry opportunities and raise and answer questions in relation to changes in nature's patterns, phonological changes or seasonal variations that are occurring because of climate change. This should be achieved while also promoting a physical and emotional connection with nature. In Year 4, children are taught to 'recognise that environments can change and that this may sometimes pose dangers to living things' (Department for Education, 2013b, p. 155), which in a SCCB curriculum also includes humans. The Year 4 guidance continues, suggesting the exploration of examples of human impact, both negative and positive, giving such examples not only of deforestation and pollution but also of more hopeful investigations into the impact of nature reserves, ponds and planned ecoparks. At upper Key Stage 2, the 'working scientifically' programme of study strongly supports an SCCB curriculum, providing opportunities for children to plan a variety of scientific enquiries, specifically including identifying patterns found in the natural environment. For example, in Year 5 children are required to observe changing life cycles of a variety of living things, hypothesising about similarities and differences and possible reasons for changes. While in Year 6, the focus is on adaption and evolution to develop this learning.

However, a broad SCCB curriculum ranges beyond science. English is both a subject and a medium for teaching across the whole curriculum, providing opportunities across all genres to develop spoken word, reading and writing knowledge and skills. Including SCCB in the English curriculum provides opportunities to develop spoken language in the form of informed discussions. Climate change facts can be used to stimulate performance and persuasive writing. Pupils could write scripts for a news item or documentary, or write to local governance to consider the issue of local pesticide use, tree felling or re-wilding areas. Such tasks explore writing forms while demonstrating vocabulary and grammar knowledge/understanding.

Case study

Mixed Year 3/4 English

A lower Key Stage class with mixed Year 3 and 4 children took part in a role-play activity as part of wider learning and discussion into a declining bee population in the area. This took place following a citizen science survey of the possible impact of climate change on pollinators in a school garden. Children took on a variety of roles, including a bee, beekeeper, habitat conservationist, keen gardener, scientist, farmer, local councillor (who chaired the debate), and school councillors tasked with developing a school garden to promote nature and biodiversity.

Before the discussion, they watched videos and conducted research online to help them maintain a point of view that was well-supported with facts while they were debating. This provided a better understanding of the perspectives of others and what action might be taken. This led to a class assembly, speaking to a whole-school wider audience, as well as writing class letters to a local MP asking them to lead action for change. With an emphasis on English grammar, the process explored the use of modal verbs and adverbs to indicate degrees of possibility (will, would, can, could, may, might, shall, should, must and ought). The word 'environment' was also included in the Year 5 and 6 word list for English.

In **maths** or **science**, a SCCB curriculum focus provides opportunities to explore larger numbers, measures, geometry and data. As part of a wider investigation into habitat loss, for example, children might model and graph sea levels rising or measure and map areas of ice melting. They might recall, use and apply mental methods to derive facts about global warming or ice loss. They might calculate energy-saving measures or the carbon capture and footprint of a school site as part of a sustainable planting project.

Case study

Year 4 Maths

A Year 4 class collected number data about specific animals (ants in a paved area, worms in a set-aside earth area, flying insects in a planted area, birds near a tree area), recording temperature and weather (including percentage of cloud cover) in a monthly survey and noticed patterns in the outdoor area as part of a school site survey. They worked with Year 5 maths buddies to explain what they had found, discussing which was the most effective way to present the data and why. Some children began to make hypotheses and draw conclusions from the data, with some making predictions to test in a further investigation.

Case study

Year 3 Science

A Year 3 class recorded rainfall and temperatures over time in a monthly data collection activity over the course of a school year. They then compared this with historic data conducted the previous year and presented the data using bar charts. They looked for similarities and differences, as part of wider learning into the differences between weather, climate and climate change, while being involved in a longer-term citizen science project.

Case study

Year 6 cross-curricular project

In a cross-curricular Design and Technology project to create a bee hotel and sustainable bee-friendly growing area, Year 6 children weighed seeds, compost and clay powder to create seed bombs. They measured the water volume needed to nurture and water the seedlings. They recorded the temperature in the greenhouse (although some were grown on a windowsill of the classroom as part of a wider experiment into growing conditions). They calculated the cost of trays of seedlings and the cost of each seed bomb and sold these to parents to fund materials for a raised bed area, measuring lengths of wood and calculating surface area and volume of soil compost needed to fill the beds.

Case study

Year 4 Science

As a part of a wider science curriculum looking at properties and changes of materials, Year 4 children explored solutions and filtering. They devised experiments to separate salt from water and use saltwater solutions to investigate how salt affects the density of water, as part of an exploration into the impact of melting ice on the oceans. Then, as part of a focus on states of matter and inspired by a puddle in the playground, they recorded evaporation over a period of time and hypothesised the impact on local wildlife.

In **art**, a SCCB curriculum provides many opportunities for illustration, persuasive images and posters. Children can draw diagrams of the biodiversity of the school site, recorded from direct observation in sketchbooks and nature journals. They can study the effect of climate change over time by drawing, painting, recording or photographing the school setting to capture local changes over time.

Geography offers the opportunity for experiential place-based learning, including fieldwork and studies of locational changes and human impact to 'deepen understanding of the interaction between physical and human processes' (Department for Education, 2013, p. 198) and environments, and provides an opportunity to reinforce the difference between weather and climate. Orienteering on the school grounds using map work to find specific locations or to complete tasks at control points not only uses geographic skills but is also fun! Orienteering offers the potential to be an active medium for cross-curricular learning across the whole of the Key Stage, as exemplified in Bloomsbury's *National Curriculum Outdoors*; a complete scheme of work series (Lambert, Roberts and Waite, 2020).

Music outdoors provides an ideal location for freedom of expression, sensory development and creativity. Children can perform their own musical compositions on the school grounds, using natural materials and instruments, tuning into nature. Perhaps inspired by the work of drummer and ice music composer Terje Isungset, or Jana Winderen and Andreas Bick, who create collages of ice sounds recorded across the Arctic, children might record environmental sounds and use these to compose 'soundscapes' (Lambert, Roberts and Waite, 2020). Recording a moment in time creates a sound 'time capsule' as a way of capturing evidence of changes in the environment over time.

There will be parallels with other national statutory curricula:

- The Curriculum for Wales (2022) acknowledges science and technology as drivers for change in society. It emphasises the importance of scientific and technological literacy in the modern world, encouraging an enquiry-led approach where children are encouraged to be curious and to search for answers.

- The Scotland Curriculum for Excellence (2004) includes the 'Planet Earth' science experience and outcomes. This encourages children to explore and measure the weather, explore climate zones around the world, compare and describe how climate affects living things and contrast weather and climate conditions across Britain, discussing the impact on living things.

- In Northern Ireland, the interdependence of people, plants, animals and the environment is an enquiry-led cross-curricular focus. This is led by the key question, 'How do living things interact with each other in the environment?' (CCEA, 2007, p.90). SCCB is defined in Geography, investigating some of the ways people affect or conserve the environment both locally and globally, what action might be taken on a local or global issue, local litter issues and improving or preserving local habitats. SCCB is also utilised in History by exploring local and global impacts of the use of natural resources through time. Furthermore, it includes a personal development and mutual understanding aspect of connection to the environment, with a 'taking action' approach to environmental issues by 'playing an active and meaningful part in the life of a community' and being concerned about the wider environment' (CCEA, 2007, p. 99).

Within existing statutory curricula, the emphasis can be shifted to focus on SCCB not only to raise awareness of the issues but also to take action for change. Detailed examples of how National Curriculum-linked SCCB knowledge can be delivered in an outdoor setting are found in Chapters 4 and 5. These links can also map successfully to curricula elsewhere in the world.

7 Sustained impact and measuring impact

It is important to measure the impact that the cycle illustrated by figure 2.1 has had on the overall delivery of outdoor learning. It is vital that there is ongoing evaluation in line with regular school leadership policy. In the case of climate change education, this might look like improvements in the biodiversity richness of the school site (more examples of how you can develop biodiversity on your school site can be found in Chapters 4 and 5). We measure the impact of outdoor learning to evidence what difference is being made to a child's engagement, attainment, wellbeing and achievement.

To assess the impact, you need to start with baseline data obtained through qualitative and quantitative measurements. These might be questionnaires, attendance data, behaviour incidents data and/or SATs data. If you are aware of the baseline data, you can then measure the difference a programme has made within the school. It is important to include all stakeholders in monitoring the impact of a programme to ensure continuous improvement and buy-in to future developments.

Case study

Measuring the impact of outdoor learning on pupil outcomes

The Arena Wild Tribe Explorers 6- or 12-week intervention programme aims to build self-esteem for pupils identified as requiring support in developing their behaviour for learning. The programme uses emotional coaching to support the delivery of structured social, emotional and mental health (SEMH) outdoor sessions, while also upskilling school staff using this approach in the delivery of outdoor learning.

The assessment tools provide a baseline, alongside post-programme evaluation – a self-esteem indicator questionnaire and strengths and difficulties questionnaire – with input both from the children and the teachers. As a result of the self-esteem indicator, an overall score is produced based on Morris's (2014) categories:

- Sense of Self – having a good idea about who you are e.g. knowing and being comfortable with your likes, dislikes, strengths, vulnerabilities, preferences, temperament, feelings and needs.

- Sense of Belonging – how aware of and comfortable you are within a setting or relationships with other people.

- Sense of Power – an inner knowledge of your ability to have an impact on the world around you.

The programme provides a set format for each session, such as sharing food and drink, open talk time, shared activities, a creative activity and a physical activity, but it is important to emphasise that the planning is tailored to the SEMH needs identified by the data, age and prior experience of the children.

An example of the impact of this programme was evidenced in a group of 18 children who took part in the 6-week Explorers programme. Out of the 18 children attending, eight children had a vulnerable score (one with a particularly low score). Sense of personal power was the lowest score in 16 out of 18 children.

As a result of the Explorers programme, data from the strengths and difficulties results and pupil conferencing showed:

- 100% of children had an improvement in their overall score.

- Sense of personal power increased significantly in all children.

- Only two children retained a vulnerable score as a result of the programme, with significant improvements seen in these children.

- Children reported feeling more confident, were able to work with others, could lead tasks, felt more valued and took more pride in their work.

Further noticeable impact as a result of the programme, reported by the school, included:

- Teaching staff to create and demonstrate a 'self-esteem building' environment.

- Raising self-esteem (evidenced by the self-esteem indicator).

- Improved attainment and attendance.

- Stronger relationships with peers and adults.

- Better engagement with their own learning and participation in class (inclusion).

- More resilience and the ability to deal with their own challenges.

- Improved strategies for managing their thoughts and feelings.

> 'The Wild Tribe Explorers programme has developed the children's self-confidence and self-awareness. As a result of the programme, many children are willing to try new things outside and inside the classroom. The impact of the programme has been significant and has also supported me with evidence in my SEF [School Self Evaluation Form] and School Improvement Plan.'

A. Bassett, headteacher, July 2019

Reflection

What next?

- How can learning be evidenced and recorded in the outdoors?

- How can existing strategies and systems to gather evidence that are working well in your setting be applied to outdoor and climate change learning to monitor and evaluate impact and progress?

- What baseline assessment methods could you put in place to measure the impact of outdoor learning?

- What SMART outcomes could you use?

- How will you gauge the quality of the data and how will you use it?

- What is the purpose of gathering evidence of impact and progress? To assess specific and measurable elements, such as curriculum objectives. But don't forget the hidden learning that will take place! This includes:

  - natural connections

  - speaking and listening

  - self-confidence

  - resilience and independence

  - modified behaviour

  - increased vocabulary

  - concentration

  - dexterity, tool use and gross/fine motor skills development

  - environmental awareness

  - wellbeing levels suggested at the start of each unit and linked to PSHE.

Concluding thoughts

Sustainable leadership and a whole-school approach

Adopting a whole-school approach, underpinned with effective and sustainable leadership, encourages a collaborative and supportive approach to outdoor learning and climate change education. It creates a shared understanding of outdoor learning purpose through a shared knowledge-based vision. This supports the whole-school community to acquire specific knowledge and skills to support the safe delivery of outdoor learning as part of effective teaching practice, and provides a framework to support planning for consistency across settings and Key Stages.

Reflection

How can you do the following things in your school?

- Apply the knowledge and skills of existing practitioners, strategies and systems to outdoor learning approaches that are already working well in your setting.

- Align procedures, codes of conduct, assessments and learning outcomes to existing practice to provide an evidence-informed approach.

- Regularly (informally and formally) share, monitor and review outdoor learning as it becomes more established in your setting.

- Share and celebrate successes as part of the data-gathering process.

- Apply existing learning outcomes and systems for assessment to provide an informed approach to outdoor learning practice to raise attainment.

- Ensure progression of knowledge and skills across the Key Stages.

- Feel confident about taking an outdoor learning approach.

- Ask for support when YOU feel you need it.

- Assess the outdoor spaces and areas you have to deliver cross-curriculum learning outdoors in your own setting. What resource does each area offer to explore the curriculum? How could evidence of progress be recorded in this setting?

Key messages

- Establish your ethos, mindset and vision for sustainable outdoor learning.

- Develop your whole-school approach by auditing existing provision as a baseline.

- Create a 3-year sustainable site plan.

- Upskill your staff by providing continuous training.

- Consider the role of your pupils and other stakeholders.

- Develop a sustainability and climate change curriculum across Key Stages which meets the needs of your school.

- Evaluate the impact of your whole-school plan.

Chapter 3
Creating the natural infrastructure of nearby nature on school grounds

Introduction

A common challenge for schools wishing to offer children regular, nature-based learning is finding time to visit natural environments if these are far from the school site. The shorter the distance is to reach a suitable green space, the more likely it is to be used for teaching and learning (Waite et al., 2016). This chapter will help you to green your school grounds by offering guidance about how this can contribute to sustainability through climate change mitigation and biodiversity. It begins with a rationale for why this is important and some background and reflection on guidelines for school grounds. It then looks at how these spaces can help mitigate climate change effects and promote sustainable practices. Three case studies provide vivid illustrations of how some schools are addressing designing these spaces with children's input and the support of external organisations. Resources to help your school navigate the process are provided within the chapter, in appendices and through relevant web links. The suggestions and examples of ways that school grounds can be enhanced are intended to help schools provide rich learning environments, oases of biodiversity and support the growing need to teach sustainability and climate change (Department for Education, 2022; Osterloff, n.d.).

Grounds for sustainability and climate change mitigation

In the Natural Connections Project, which worked with 125 schools across the southwest of England, many schools began their journey of embedding teaching outdoors throughout the curriculum by improving their school grounds to provide nature nearby (Waite et al., 2016). Nearby nature refers to natural environments within a 10-minute walk. Earlier research had shown that the surroundings of schools were sometimes designed more for visual appeal and tidiness than their learning potential and school rules sometimes dictated that there should be no access to the playing field for three days after rain (Waite, Davis and Brown, 2006). In a wet corner of the UK, such a rule severely restricted how often children could use the grassy areas of their grounds! Yet, there has been some pressure in recent years to not only see large grounds as a valuable resource to improve the condition and sustainability of schools, but also to provide housing and new community spaces (Whittaker, 2022). Caution is needed to ensure that the space for potentially precious miniature nature reserves in school sites is not lost in the process. The National Education Nature Park is part of the Government's Sustainability and Climate Change Strategy or SCCS (2022), which is intended to ensure that their potential contribution to sustainability is realised (Natural History Museum, n.d.-c; Royal Horticultural Society, 2022). It also seeks to bring natural features onto school grounds that are short of outdoor space to fulfil ecosystem services functions.

This includes climate change mitigation, sustainable food production and biodiverse habitats, as well as supporting children's access to nature, environmental awareness and wellbeing. The Department for Education's non-statutory guidelines for the use of space on school grounds dating from 2014 are shown in Table 1 below. If school grounds were compartmentalised in their practical use, as suggested by the current guidance, the natural environment would be practically overlooked in the allocation of space. The suggested minimum site allowance for habitat has dropped from 200m² in the Department for Education guidance in 2004 to zero in the currently applicable guidance. Also, the per pupil allowance of space is reduced by three-quarters from 1.5 to 0.5 m².

Types of outdoor provision	Minimum site areas in m²	Area for each KS2 pupil in m²
Soft outdoor PE	–	35
Hard outdoor PE	400	1.5
Soft informal & social area	600	2
Hard informal & social area	200	1
Habitat	0	0.5
'Float'*	600	5
Total net area	2000	45

Table 3.1: Government guidance for outdoor space on primary school sites.

*Float is space that can follow the distinct priorities of schools, as there is usually a difference between the total area for specific purposes and recommended total net area.

(Department for Education, 2014)

Government guidance also sets out the rationale and suggestions for provision within the school grounds, arguing for a design that is responsive to curriculum learning, holistic development and cultural needs of pupils both in and out of class time in its informal social areas. It advises that soft play areas should be safe and provide shade, and include grass to sit and socialise on and sloped grass banks forming natural amphitheatres. Habitat areas, it is argued in the 2014 guidance, 'can include a range of outdoor classroom spaces and designs, to provide a valuable resource for teaching and learning across the whole curriculum. The total habitat area should include grounds developed for a range of supervised activities, for instance, meadowland, wildlife habitats (such as ponds), gardens and outdoor science areas to support the curriculum.' (Department for Education, 2014, p. 58).

In earlier guidance leading up to the Sustainable Schools Initiative (Department for Education and Skills, 2006b), it was also acknowledged that 'increasingly, such spaces are being given a central and accessible location. Landscape design has great potential for promoting a sense of ownership of space by pupils and staff, thereby encouraging people to take greater care of their surroundings. However, some wildlife areas should normally be undisturbed, so are best positioned away from busy social areas. Such areas usually need to be fenced off, both to protect the habitat and for the safety of pupils.' (Department for Education and Skills, 2004, p.59).

Unfortunately, the more recent guidance (Department for Education, 2014), does not make the case as strongly for habitat areas. Furthermore, soft, informal and social areas that might additionally provide wildlife sanctuaries and climate change mitigation are also given lower priority than hard surfaces outside foundation stage classes and PE.

One possible reason for the apparent dilution of sustainability messages may be that, despite the fact that a Sustainable Schools Initiative (Department for Children, Schools and Families, 2008) was launched nearly twenty years ago, progress towards sustainability has been as slow within the education sector as in wider society. Competing priorities and changes in government may have overtaken this vitally important ambition. Societal attitudes, however, seem to be turning again. For example, the COVID-19 pandemic highlighted how many people appreciate the natural environment as a source of personal wellbeing and, as discussed in Chapter 1, young people are becoming increasingly activist in global sustainability matters. Issues of sustainability that were pressing 20 years ago are now even more so.

While overuse can degrade small wild areas by eroding plants and disturbing wildlife, separating children from nature is artificial – humans are, after all, part of nature. Indeed, it may be harmful for both our children and the natural world to set up barriers between them, for, if young people do not come into regular contact with nature, how will they learn to love it and care for it (Lumber, Richardson and Sheffield, 2017)? There is now a wealth of research that shows being outside in green and blue spaces is beneficial for human health and wellbeing. Mathew White and colleagues published a study (2019) which showed that just 120 minutes per week is enough to improve our wellbeing, while Alice Goodenough and Sue Waite and others (Waite et al., 2016) found that early contact with nature shaped future pro-environmental attitudes and action. The main message from this inconsistency is that outdoor space guidance is lagging behind renewed government initiatives and popular demand for sustainability. However, ensuring mixed purposing of zones, which are currently separated in the guidelines, can help to

maximise space for nature. Furthermore, schools with nature on their grounds can provide universal access for children to these multiple benefits.

As discussed in Chapter 1, the UK Government's Children and Nature programme, COP 26 and the regular Monitoring Engagement with the Natural Environment surveys of visits to nature by children (Natural England, 2019) have provided a refreshed recognition that nature needs to be made accessible, understood and cared for in the face of climate change and sustainability challenges. Part of this renewed drive was delivered through the Nature Friendly Schools programme led by The Wildlife Trusts between 2019 and 2022 (The Nature Friendly Schools, n.d.). It provided transformational outdoor learning opportunities to more than 46,000 children in England and was funded by the Department for Education, Department for Environment, Food and Rural Affairs and Natural England. The Wildlife Trusts led a consortium of voluntary organisations, namely Groundwork, Young Minds, the Field Studies Council and the Sensory Trust. These organisations worked with teachers from schools known to have high levels of pupils eligible for pupil premium (an average of 46% eligibility across all participating schools) and significant numbers of pupils with English as an additional language (18% of the pupil population of participating schools).

During the project, 1,870 teachers from 187 schools were supported to:

- Make their school grounds more natural, and better as an outdoor educational resource.

- Make more (and more effective) use of off-site natural places.

- Deliver a range of high-quality outdoor activities intended to benefit children's mental health, wellbeing and engagement with learning.

- Collect evidence to support the evaluation of the project.

On completion of the project, 94% of participating schools reported that it had increased pupils' care and concern for the environment, interest in learning and personal resilience. 97% said it had increased pupils' interest in the natural world and helped them to develop new skills. A substantial proportion of participating schools also felt that the project had delivered significant benefits to teachers, including increased confidence to lead lessons outdoors (89%), greater confidence to increase the range of subjects taught outdoors (85%), increased confidence and willingness to try new things (87%) and increased ability to translate curriculum into outdoor spaces (83%).

These figures illustrate the marvellous news that it isn't an either-or situation of ensuring benefits for nature or children. We can develop our school grounds to be brilliant and exciting learning environments while also enriching their biodiversity (Natural History Museum, n.d.-a) and provide the natural opportunities that this combination brings for other species and children's wellbeing.

However, we should reiterate that greening school grounds to meet climate change challenges is by no means a new situation, despite the urgent need to achieve this. Coinciding with the Sustainable Schools Initiative (Department for Education and Skills, 2006b), in their guidance on Designing School Grounds published nearly twenty years ago, the Department for Education and Skills suggested:

'Climate change can no longer be ignored and should be considered in all school grounds decisions. This means:

Planning for a drier climate, hotter summers and potentially wetter, warmer winters.

Using plants that best suit these conditions – drought-resistant (often low-maintenance).

Thinking about water table changes and how to innovate to help collect much-needed water for growing plants.

Planning bog gardens and ponds to take advantage of retaining water.

Using sheltered south facing walls for Mediterranean-type vegetation.'

(Department for Education and Skills, 2006b, p. 88)

Yet, the provision of these elements on school grounds is currently extremely patchy, perhaps because of barriers such as lack of funding, expertise, and confidence to rewild and develop areas of the school site to meet this challenge. (There are examples of some ways that schools are tackling this. For links to Landscape Strategy in the curriculum, visit droxfordjunior.co.uk. For more information about projects with the Aquifer Partnership, visit wearetap.org.uk.)

Encouragingly, the latest impetus within the Government's SCCS (2022) builds on the Nature Friendly Schools programme to recognise the role of education in combatting climate change effects. It is driven through a partnership led by the Natural History Museum with the Royal Horticultural Society, supported by the Royal Society, the Royal Geographical Society (with IBG), Manchester Metropolitan University, Learning Through Landscapes, UK Centre for Ecology and Hydrology and the National Biodiversity Network Trust. These organisations, and others with expertise in how to achieve these goals, will support schools and their pupils to lead and manage their own school-based Nature Park as part of the National Education Nature Park programme, enabling students to:

- Model the staff of a National Park or nature reserve as managers, ecologists, communicators, fundraisers, grounds people and data analysts, and gain a systemic and situated understanding of how to sustain ecosystem functionality.

- Choose, plan, and implement evidence-based site improvements and habitat enhancements, and monitor biodiversity gains over time, thus putting into action enhanced knowledge and understanding of challenges and rewards associated with creating more biodiverse school sites.

Through these actions involving situated experiential learning, which encourages both cognitive and affective engagement, it is anticipated that more sustained change in attitudes and behaviour will be achieved (Waite and Pratt, 2015). The partnership is also in the process of developing free, curriculum-based climate education resources, lesson plans, and schemes of work (see their website: https://www.educationnaturepark. org.uk/) so that green infrastructure and appropriate pedagogy will be mutually supportive. Chapters 4 and 5 of this book provide detailed lesson plans that will complement the development of school grounds' potential for learning, climate change mitigation and biodiversity.

Climate change mitigation on school grounds

Planners are increasingly recognising that greener school sites can cope better than hard-surfaced playgrounds with extremes of weather linked to climate change, such as excessively hot summers and flooding caused by storms. While other parts of the world, such as Australia, have been at the forefront of such adverse effects (Pfautsch, Wujeska-Klause and Walters, 2022), it is inevitable that many problems identified in these countries will become more common in the UK in the future.

Children have been shown to be more sensitive than adults to heat stress and be more susceptible to heat exhaustion and burns directly from the Sun or through heated surfaces. Where no shade is provided over hard-surfaced areas, black tarmac and metal features absorb and retain solar warmth and can burn children's sensitive skin (Antoniadis et al., 2020). According to Pfautsch and colleagues (2022), tree planting is preferable to fabricated structures as a long-term cooling strategy in playgrounds. A tree's wide leafy canopy of dense foliage not only provides shade but also helps cool the air around them through transpiration.

Urban playgrounds form heat islands that are hotter than surrounding streets in periods of extremely hot weather that are becoming increasingly familiar in the UK. Plants, especially trees, can significantly alter the microclimate on school grounds. Nature-based solutions are widely accepted as effective ways to mitigate these issues. For example, in Paris, the 'Oasis' project is planting 700 schoolyards (about 197 acres) with trees and vegetation, thereby increasing green space in the city by 20% and reducing the levels of heat within, and 200m around, these sites. In Holland, the Amsterdam Impulse Schools programme is intended to promote active play, contribute to rainwater absorption, support outdoor educational activities, and encourage citizen participation in the promotion of sustainability by adding 25% more green areas to participating schools (Antoniadis et al., 2020). These programmes show that greening school grounds can contribute not only to the wellbeing of schools' own population but also make a difference to the wider community's environment.

Demonstrating sustainability

Schools teaching about climate change and sustainability face a major challenge if they do not demonstrate such concepts in real life. Embodied experiences in school grounds accompanied by careful contingent support from the teacher are highly effective pedagogical strategies (Green & Rayner, 2020). We know that many children in the UK do not know where much of their food comes from or the effect of changes in weather patterns on crops and that the link between growing, harvesting and consuming has been lost (Food A Fact Of Life, n.d.-a). Furthermore, one in eight households have no private outdoor space, with Black people nearly four times as likely as White people to have no access to outdoor space at home, whether it be a private or shared garden, patio, or balcony (37% compared to 10%) (Office for National Statistics, 2020). These statistics illustrate some stark contrasts in access to nature that exist in relation to both socioeconomic circumstances and a lack of cultural inclusivity in some nature-based organisations (Waite et al., 2021). School gardens therefore can offer a chance to redress some of that inequality by reconnecting **all** pupils with the cycles needed to sustain food production and slow biodiversity loss. They can help them to understand the threat to crops when weather events disrupt cycles. The Food A Fact of Life programme, which was originally launched back in 1991, continues to provide evidence-based advice to schools. They have also produced a guide to 'growing clubs', which give children first-hand experience of growing food on their school site, whether in raised beds and pots on tarmacked playgrounds or through a designated garden area (Food A Fact Of Life, n.d.-b).

As we have seen in Chapter 1, much media attention focuses on environmental disasters that climate change is causing, but less information is shared that can give children hope to effect positive changes. Involving children in discussions and creative action at a local level with changes to their school grounds may help reduce a tendency towards climate anxiety and helplessness (Waite et al., 2016; Rousell and Cutter-Mackenzie, 2020).

The latter authors argue that climate change education should respond to 'the existing beliefs, attitudes and situational contexts of specific audiences, rather than focusing on what people don't know or understand about climate change' (Rousell and Cutter-Mackenzie, 2020, p. 203). In Appendix 5, you can find an envisioning tool to help elicit children's views on what they would like to include on their school grounds to help nature. Hickman et al. (2021) found 92% of UK children in their survey were worried about climate change and over three quarters attributed blame to governments for not doing enough soon enough to avoid its worst effects. Local engagement and measures described here that enable positive changes in schools, of course, need to be mirrored by urgent action at a national and global level to address children's understandable anxiety.

Greta Thunberg in Sweden was one of the first young people to express their concern publicly and forcefully that climate change was not being taken seriously enough, withdrawing from school every Friday to protest the lack of attention to this topic in her education. More recently, she gathered contributions from eminent thinkers in a book that provides valuable and accessible foundational knowledge about climate change (Thunberg, 2022). The case study on the next page describes a series of Swedish studies that sought to inform and support kindergartens and schools in developing sustainable school grounds.

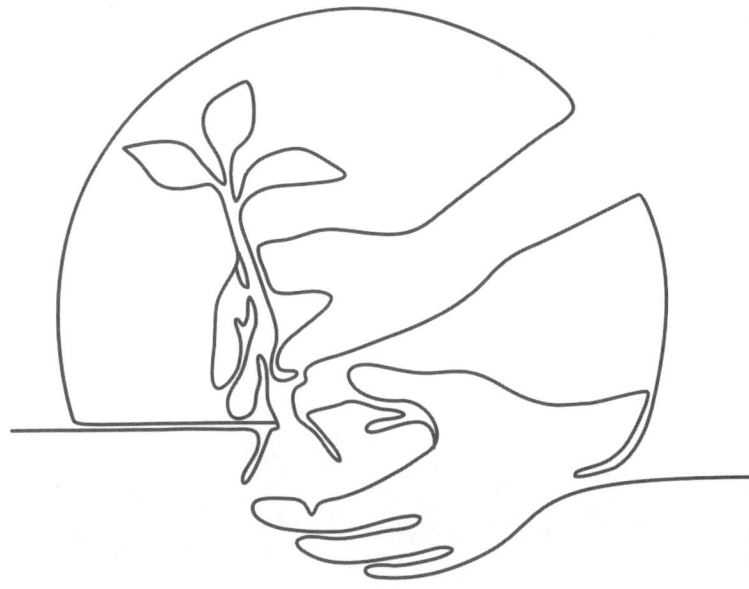

Case study

Strengthening ecosystem services in Swedish preschool and school grounds

The process of strengthening ecosystem services – nature's contributions to human wellbeing – on school grounds can serve as a catalyst for engaging children and adults in inclusive, health-promoting actions and more sustainable and biodiverse environments. This is the experience from a collaborative research and development project in Sweden, where knowledge was developed to improve outdoor environments for preschools over time, and in which children's play and learning were promoted by their active participation. The research and communication project, 'Multifunctional urban outdoor environments that promote health and sustainability: a study of collaborative processes for strengthened ecosystem services', was supported by Formas from 2016 to 2020. The project was carried out in collaboration with the Jönköping County Public Health Department, ten preschools and municipalities in the region Jönköpings län. Ten preschools were invited to participate in the two-year project. In addition to the preschools' own practical work on their grounds, the project included several meetings to share plans, progress, challenges, experiences, and inspiration between the preschools.

Some ecosystem services innovations proved more popular than others with staff and children. Planting nectar-producing plants and constructing insect hotels to support pollination were most often developed, although plantations yielding fruits, berries, and other edible plants (while supporting insects with nectar and pollen) were almost equally popular. Children enjoyed helping staff produce bird boxes, often fancifully painting them in bright colours. Some preschools put up bat boxes as well. Mini forest gardens (a form of permaculture that produces edible leaves from perennials, berries and fruits under conditions that require little effort in terms of watering and weeding after initial set-up) were created by some preschools. These are very suitable for school grounds where ongoing maintenance, especially during holidays, can be problematic.

Less commonly adopted ecosystem service innovations included constructing small ditches or 'wetlands' to catch surface water runoff and improve the microclimate, for example, through pergolas to reduce harmful UV radiation. A few schools installed child-safe vessels to collect rainwater from their roofs for watering.

Obstacles to the adoption of innovations were a lack of participation and knowledge about sustainability and ecosystem services, and a lack of communication between preschool staff and other staff. In one case, the janitors trimmed down pumpkin plants that children were growing and cleared away organic matter that could have been composted. We also noted that sometimes children did not understand the purpose of certain ecosystem service innovations, such as insect hotels. This may have been because they had not been fully included in the process of installing them. An important conclusion from this project is therefore that children, alongside the whole-school community, must be involved in the projects from the start.

We have followed this up with a communication project intended to facilitate the broad participation of preschools and schools in the transition to sustainable local environments. These were to have strengthened ecosystem services, and good play and learning environments, through supporting collaborative processes that promote action competence and learning for sustainability. As a first step, our results and participants' experiences have been disseminated through webinars, films, infographics and tutorials to preschool teachers, school teachers, children, janitors, kitchen staff, school principals, technical administration and housing companies, and other target groups throughout Sweden on hsr.se/smulle. Keep Sweden Tidy's well-established and extensive school and municipal network is the main hub for communication. These same target groups have also been invited to submit their own films and case studies in different categories: Learning activities; Ecosystem services; Collaborative processes – obstacles and opportunities. Our hope is that this network will stimulate further growth of these practices.

Articles related to this case study are listed in the further reading section at the back of the book.

Further images of the developed school sites and associated reflection points are available on the Bloomsbury Education website.

Figure 3.1: Soil, trees, edible plants, and pathways: ingredients for a mini forest garden.

Ellen Almers, Per Askerlund,* Sofia Kjellström, Robert Lecusay & Tobias Samuelsson, Jönköping University*
**Authors of this case study within this research team (the rest comprise the research team)*

What this case study clearly demonstrates is that the inclusion and participation of children at every stage of development and management of school grounds are vital if they are to recognise the innovations' importance, echoing an earlier point that one needs to have contact with and care for nature to understand, value and look after it. Features that support teaching of climate change and sustainability were successfully blended with play and other learning opportunities in these settings. Establishing a network that combines sharing ideas with peers and expert advice has also supported schools in their endeavours.

Reflection

Think about how the above applies to your school:

What mechanisms for gathering pupils' opinions do you have in your school?

How accessible is a green space to your school?

How often do the children have opportunities to play and learn in natural surroundings?

What local nature-based organisations might be able to assist with developing your plans for sustainable school grounds?

Designing and developing natural learning environments

It is important to start any process of change by evaluating where you are now. Every school is different, and each school site will have a variety of strengths and challenges in making it more suitable for addressing climate change and sustainability.

Reflection

Using Figure CS1.3 (online) 🔑: 'Some rich possibilities for greener school grounds', either as a self-reflection, staff development or class activity, consider these questions associated with the image:

• How many green innovations can you or your pupils spot in the picture?

• How many of these could you work towards in your school?

Some common principles and guidelines will help guide your thinking and plans for your specific situation (Department for Education and Skills, 2006b; Department for Children, Schools and Families, 2008; Evergreen, 2011). This section will introduce these principles, how to get started, then focus on nature, learning and safety. There are also lots of ideas on the Natural Learning Initiative website that can support you in gradually adding green features to your school grounds too.

Principles

• Know your space. Conduct an audit and survey, mapping the current zones and features of the site. (See Appendix 4 for a brief audit tool for sustainability status of grounds.) Think about the Sun's movement through the day and seasons to plan shade provision and consider where stormwater run-off could usefully be channeled.

• Make your focus on how to enhance the natural environment in your school site and make decisions that will be manageable and sustainable in the future. Consider how your actions will help nature be restored in your space and how they build on the unique existing environmental features of your site. For example, take into account where trees are currently sited and keep as many mature specimens as you can to provide shade and habitat.

• Whenever possible, recycle and repurpose existing features. For example, you could make pathways from wood chippings, use old buckets and wood to make containers and raised beds for gardening or plant trees into carefully prepared holes in hard landscaping to provide shade and/or drainage in these 'deserts.'

• If you need to buy new things, choose high-quality natural materials that will last well whatever the weather in areas of heavy traffic. Make sure they will be relatively easy to maintain and are sourced ethically.

Getting started

• Consult existing school and community groups, ensuring mixed representation of pupils, teachers, other staff, governors, parents, the local community and natural environmental organisations. Set up a school grounds management group from these partners, ensuring a member of senior management is involved and the process of design and management continues to be participatory (Austin, 2022). Also, ensure that the group is of mixed ages, so all can benefit from sharing different levels of understanding and experience within a culture of care for others, including plants (Lloyd and Paige, 2022).

• Make sure that children's developmental needs for play and learning are integrated by including different zones in your plans and considering your school's distinctive cultural and ecological identity. Austin (2022), studying eight school gardens in Ireland, found considerable diversity in their size, content and use, but noted a shared emphasis on food growing and environmental awareness. She concluded that schools find a way to use their garden spaces in a way that works for them and their community.

• Embed project management of the proposed changes in the school improvement strategy and devise a detailed action plan with an allocated budget. Be mindful that plans may evolve over time and regularly review them.

• Use local suppliers and contractors whenever possible to reduce the carbon footprint.

Focus on nature

- Using your school ground's zoned map, reflect on space allocation for different purposes and consider if more areas could combine natural and educational benefits.

- Conduct habitat surveys to establish what currently exists (see Chapter 2 for how pupils can play a major role in this). Make sure you consider not only the number of trees or animals but also their locations across the site. This is so sun and wind protection are taken into account and the survey supports children's growing environmental awareness.

- Consider what habitats are missing (meadowland, woodland, fruit and vegetable gardens, ponds, deadwood, etc.) that would enhance the site and enrich the biodiversity. For example, planting hedgerow plants around boundaries or as links between greened areas can create rich wildlife sanctuaries and help to connect different wild areas of the grounds, helping wildlife move between them safely. Native trees and vegetation are preferable to ornamental shrubs and mown lawns as they harbour many more indigenous species and thus will offer valuable environmental learning opportunities. See Appendix 6 and Appendix 7 for a list of UK trees for shade and a list of suitable plants, or visit the Royal Horticultural Society school gardening website.

- Think about the size of wild areas to ensure that they are big enough to be sustainable for their intended educational and wildlife uses. Coppicing (cutting back a tree's main trunk every few years to promote multiple smaller shoots) can help keep trees manageable in smaller spaces and provide woody material for projects.

- Create well-connected durable paths that will reduce footfall on sensitive and/or small wild areas.

- Make sure rainwater is collected in securely lidded water butts to water your plants. Alternatively, direct surface water run-off into ditches to form mini wetland habitats.

- Plan low-intervention management regimes of grounds so that trees, grass and wildflowers can seed and regenerate, saving energy and money.

- Compost any waste organic material and use the compost to improve your soil. At the same time, ensure that children and staff remove and recycle rubbish that might harm wildlife.

- Consider introducing manmade features to boost wildlife habitat, such as bird boxes or bug hotels in suitable areas protected from excessive sun or rainfall.

Reflection

It is important to recognise the opportunities and challenges of your own particular site.

Create a list of these ideas in three columns: now, later and never. This will help you consider where the quick wins might be, what needs to be worked towards over time and what is simply not feasible in your situation.

Try to be as ambitious as you can in determining which column is appropriate. Use this reflection as the basis for development planning.

Focus on learning

- Incorporate children's interests and consider any special needs of your school's population when developing themed gardens. For example, a sensory garden can help children's wellbeing by appealing to their senses and providing a calm place. Growing a range of vegetables can reflect the cultural profile of your community.

- Review habitat areas and their co-location with other zones on the school grounds for their teaching and learning possibilities (See Chapters 4 and 5 for how children can get involved in this process). What practical messages do the physical features convey about climate change mitigation and sustainable living to teach children about these matters?

- Use possible disturbance to habitats as a teaching point about how to care for nature.

- Develop flexible areas with multiple uses to stimulate children's creative engagement with the natural environment, facilitate transdisciplinary learning and avoid reinforcing a perceived division between human and non-humankind.

- Embed school policies on climate change response and sustainability in the school development plan, including curriculum-based learning and citizenship responsibilities.

- Remember student experiences in a school garden are 'supportive in connecting them to Earth, their curriculum, and the interconnectedness of the subjects they do at school – transdisciplinary learning' (Lloyd and Paige, 2022, p. 7).

- Encourage children, parents and the community to maximise informal as well as formal learning opportunities and enjoy the social aspect of improved grounds.

Focus on safety

Your goal should be to create 'safe enough' spaces. This means assessing the risks and benefits of different areas and allowing easy monitoring of any potential hazards; it does *not* mean removing all risks (Health and Safety Executive, n.d.).

The Royal Horticultural Society school gardening website (see the further reading section at the back of the book) suggests a number of things to consider. These include:

- **Soil**

  - Check with the Local Education Authority to see that there are no previous contaminants on the land.

  - Avoid using garden chemicals.

  - Guard against possible bacterial dangers in the soil through up-to-date tetanus jabs and the use of gloves or regular hand washing.

  - Check for animal waste. If fox, dog, or cat poo is accidentally touched, wash the affected part immediately as *Toxocara canis* in faeces can cause blindness.

  - Exposed soil and muddy/leafy areas and paths can get slippery in wet or icy weather, but reminding children to exercise sensible care when moving around and wearing appropriate footwear should be sufficient (rather than banning their access to such areas).

- **Plants**

  - It is important for children to learn that many plants can be harmful if treated incorrectly. For example, some toadstools and bulbs are poisonous if eaten. Other plants, such as giant hogweed, can cause nasty blisters. Even meadow grasses can irritate if a child is susceptible to hay fever. The Royal Horticultural Society school gardens website has a list of plants that it is best to avoid (Royal Horticultural Society, n.d.). However, some hazards need to be weighed against the benefits of having a diverse range of habitats. Some suitable plants are listed in Appendix 7.

- **Insects and other animals**

  - Despite steep declines in insect populations over recent years, there will be times when insects seem everywhere – especially when your class is sitting outside! Ants are particularly active in the spring, summer and autumn and can bite, joined by mosquitos and gnats in summer. The odd nip or sting is usually only administered if an insect is feeling threatened. It is worth asking parents to check their children for ticks or their bites (red, target-like areas) as these can cause serious illness if left untreated. Nonetheless, it is important that children recognise the diversity of insects that share their space and appreciate their value within the school ground's ecosystem. Guides are available to help you and the children identify some invertebrates that will thrive in healthy ecosystems (see the Royal Horticultural Society school gardening website in the further reading section at the back of this book). Larger animals such as foxes, frogs, slow worms, birds, rodents and hedgehogs may also choose to share greener school grounds. The presence of wildlife enables adults to model how to treat our fellow non-human inhabitants with respect and avoid disturbing them unduly. This can be an opportunity to demonstrate how to protect wildlife from the challenges of climate change that affect them.

- **Tools**

  - Whether pupils are whittling wood, vegetable gardening or doing scientific work, safe handling of tools should be explained and demonstrated by adults. Tools should always be stored securely. See The National Curriculum Outdoors books (Lambert et al., 2020) for further guidance on safe tool handling.

Challenges identified by Austin (2022) for schools that had developed their gardens included: maintaining it, marshalling resources and coping with an overloaded curriculum. But her study also detailed solutions that schools had found, including being able to call on parents to help when needed, connecting with local authorities and others for donations of equipment and materials, and being creative with the curriculum (ibid., p. 714).

In the next case study from Learning through Landscapes (one of the partners in the National Education Nature Park initiative), we see how acting locally can give children insight into global issues without creating anxiety, and instead contributing to their wellbeing in the wake of the COVID-19 pandemic.

Case study

My School, My Planet

In autumn 2020, 49 schools across the UK took part in the My School, My Planet pilot project. The project was led by Learning through Landscapes (LtL) and funded by the National Lottery Heritage Fund as a response to the COVID-19 pandemic. As well as supporting pupils as they returned to school after lockdown, the project had two key aims: to connect children and young people with climate change and to help them understand the environmental impact of their behaviour on their local communities and global motherlands. Since then, more and more schools across the country have taken part in My School, My Planet and found out what makes this project so special.

Pupil voice and pupil choice are key to My School, My Planet. After an introduction to the project themes of climate change, biodiversity and soils, pupils vote for the theme that they want to explore further. The impact of climate change on biodiversity is a popular theme among pupils, and this example shows how pupils learn about the topic on their school grounds, and then plan and make changes in response to what they have learned.

Pupils begin by choosing a place on their school grounds to build a meeting place – somewhere that they can make their own and use throughout the project. Using a range of materials such as tarpaulins, long sticks and poles, they design and build a shelter.

The next step is to learn more about the chosen theme of biodiversity, how it is affected by climate change and about habitats and wildlife on their own school grounds. Pupils use surveys developed by LtL and expert wildlife partners to discover the habitats and creatures already living on their grounds. They investigate the needs of pollinators or birds they hope to attract to their grounds, and the habitats and food sources required at different stages of their life cycles. They then use this information to plan changes to their grounds.

Pupils decide what changes should be made, where different elements should be located and have control of the available budget (in some cases, they might fundraise to increase this budget). The pupils are then empowered to make their planned changes a reality.

'Children selected biodiversity as their preferred topic... They're all quite eco-conscious, but it gave a real face to that. There are real live creatures here that live here on our school, in our gardens, under the ground, in the trees. They even got to experience planting plants and then having them all ripped up by foxes overnight. There's actually that kind of awareness that there's more than just humans here in this space that was a key focus in all the activities they did.'

Senior leader at a participating school.

'It made the pupils quite proud of what they were doing and that the management staff were out and taking notice and they were commenting and liking, so it encouraged them. It built the confidence and self-esteem, and it felt more like it was a whole-school approach.'

My School, My Planet trainer.

'It's so good to be doing something for the environment... it's fun... and something new and different.'

Child at a participating school.

Throughout the project, children consider their learning about local biodiversity and explore similar issues in other nations. They also discuss how the biodiversity of their local area compares to the places their families originate from. For some, this leads to an increase in pride in their motherlands and a chance to share with others more about their heritage.

Figure 3.2: Being outside planting for biodiversity locally can open our eyes to the wider world (photo courtesy of Learning through Landscapes).

"

'What I saw between the first week and the last week was people building a pride in place, a pride of where their school was and what they were doing with it, but also a pride in the places their families come from. I had one little boy... say "Miss, Miss! Did you know there's been a mouse just found in Somalia that hasn't been seen for 50 years?"... I don't think he would have shared that as a thing to be proud of from Somalia without this My School, My Planet.'

"

My School, My Planet trainer.

My School, My Planet gives children and young people more than just an opportunity to discuss the impacts of climate change — it gives them the knowledge and practical skills to make a difference. The abilities they gain through the project will support pupils as they move into adulthood, whether through their careers or a wider interest in the world around them.

Mary Jackson and colleagues, Learning through Landscapes

The My School, My Planet case study, like that from Jönköping University, highlights the importance of involving pupils in every stage of school grounds development. The children relish making choices and it encourages their sense of responsibility and sustained positive action. Because it is meaningful to them, they independently forge links with community and heritage knowledge, both new and old. Such links help to extend local learning to global awareness and understanding of issues. It is pupil engagement in action, however small, that matters most.

Reflection

Consider your distinct school population.

- What ideas and values about the natural environment do they bring to school from their home and community? Are they excited, aware, fearful or uninterested in it?

- How might your school grounds provide a door to the natural world for children who otherwise may have little access to it?

- How can you make ideas about sustainability relevant and meaningful to them?

- Do you notice any effects on pupil and staff wellbeing when they spend time outdoors?

In the next section, some ideas for schools with very small grounds and/or hard landscaped grounds are discussed.

Some solutions for urban schools

Where space is very limited to make a wild area within the grounds, stand-alone planters can introduce biodiversity by providing flowering plants for pollinating species (Natural History Museum, n.d.-b). Vertical gardens can be created with cut-off water bottles filled with compost and attached to fences. A tub of water, stocked with pond weed, or even simply pebbles, attracts wildlife (but make sure that there are ramps out of any steep-sided containers for animals to escape if they fall in). Pergolas with climbing plants and seating underneath can provide large, hard landscaping areas with cool oases of shady areas. It is also feasible in liaison with the local authority or allotment association to adopt an area within a local park or an allotment to provide additional experience and opportunities.

Reflection

If you have little green space on your school grounds, it is quite likely that the surrounding area is also lacking in access to natural environments. This makes it even more important to try to find ways to let nature in. At a staff meeting or independently, mind map some of the ways that you can help provide a greener experience for your pupils.

The next case study from The Wildlife Trusts, which led the Nature Friendly Schools programme, offers examples of how sustainable green infrastructure can benefit both people and the planet. It shows the continuing journey that start-up support by organisations like The Wildlife Trusts facilitates. Desired changes can be phased over time to make the process towards sustainability manageable, and adopting a policy of repurposing areas and materials makes ecological and economic sense. This case study also underlines that there can be many 'winners' when greening school grounds.

Although produced as part of the last government initiative for sustainable schools in 2008, the Department for Children, Schools and Families planning guide (2008) is a whole-school support package addressing a series of logical and manageable steps towards greater sustainability – it is well worth a look.

Case study

Nature Friendly Schools: Stockwell Academy, Hull

Stockwell Academy is a two-form entry primary school on the eastern edge of Hull, on a 2.5-hectare site with large areas of closely mown grass. It is a short distance from large areas of intensively managed open farmland (to the east) and the mainly industrial area adjacent to Hull's docks (to the south). Nearly half (49%) of its 370+ pupils are eligible for pupil premium.

Teachers at the school had already approached Yorkshire Wildlife Trust for help to improve their outdoor space before joining the Nature Friendly Schools project in the autumn of 2020. The project provided them with ideas, practical help, advice, guidance and training intended to improve their school grounds for wildlife and provide a resource for outdoor teaching where pupils can experience nature as a regular part of their school experience.

During the October half term, teachers made a start on delivering their plans for an outdoor classroom with trees, wildlife habitats, herb gardens and allotments. They started by repurposing grass for a bamboo den and using pallets to create a bug hotel. Plants donated by a garden centre now provide the outline for a closed-off area and deadwood logs provide extra habitat for wood-boring insects as well as seating for children. With permission, the school also raided a skip and repurposed wood, garden waste and even a bath – all providing sustainable materials for their environmental improvements.

> "

'Wood from trees that were cut down will be repurposed for the children. My hope is that they will be able to use tools to create tent pegs, sculptures, swing seats and more. The now-bare area can be cleared fully, and we can plant wildflowers to attract bees and other insects, while chippings to the side of the area can form part of the floor for our new outdoor classroom.'

> "

Mark Hemmerman, teacher (near the start of the project).

In November 2020, a Nature Friendly Schools education officer from Yorkshire Wildlife Trust led some outdoor learning training sessions with teaching staff. These left teachers feeling much more confident, comfortable and open to the idea of adding outdoor elements to learning. Teachers reflected that this would help greatly with the school's efforts to increase the amount of time that children are able to spend outdoors across all year groups.

'My class had their first 'forest school' session which was fantastic. We noted improved communication between the children when undertaking tasks and a willingness to look after the area when it was time to tidy up. We saw children who were initially shy, come out of their shells when presented with fun, small achievable tasks. We saw brilliance in the children's imaginations as they showed evidence of intuition, such as when they created a scarecrow after I told them that birds like to prey on insects. There are elements to work on, such as listening to all voices, but this was something we expected, and we believe that this will improve with frequent sessions.'

Mark Hemmerman, teacher.

By the summer term, pop-up dens were appearing amongst the trees as children explored the school grounds, enjoying being outside in the fresh air and sunshine. Teachers noted that school attendance was highest on the days when pupils had outdoor learning sessions, with some pupils appearing to gain particular benefit from increased use of outside spaces.

Figure 3.3: Children's fascination with and care for a fellow inhabitant of a greened school grounds – hello newt! (Photo courtesy of The Wildlife Trusts).

One pupil was prevented by staff from using hand tools during the autumn term because they felt he was unable to do so responsibly. By the following summer, he was using tools, dealing with fires safely and showing great leadership. His academic attainment also began to rise rapidly, with his mathematics performance rising from 'well below' in spring to 'expected' by summer. The pupil's attitude in the classroom changed, especially towards harder tasks, and this translated into improved test results. The opportunity to learn outdoors appeared to make a big contribution to his improved academic attainment, increasing his confidence and developing a love for learning.

Levels of enthusiasm after the Nature Friendly Schools programme ended have remained high among staff and pupils. They are looking forward to further physical changes around the school, including getting every class involved in planting and growing food and flowers – illustrating the value of school grounds to the recovery of nearby nature and providing useful learning environments to support the understanding of climate change, the natural world and sustainability.

Nigel Doar, Head of Science & Research, The Wildlife Trusts

Concluding thoughts

This chapter has focused mostly on plant-based solutions to meeting climate change mitigation, and how creating conditions for teaching nature-based solutions to climate change and sustainability challenges are mutually beneficial for human and non-humankind. There are other steps that schools may choose to invest in to make their site more sustainable. These include solar panels, wind turbines, reed bed filtration, permaculture, shelter belts, use of grey water, roof gardens, cycling facilities, compost bins, water butts, drought-resistant planting and traditional crafts, such as hedge laying (Department for Education and Skills, 2006b, p.89). More recent government guidance on these actions is being produced.

Any of the above will offer valuable stimuli for discussion with children about why they would help in reducing negative impacts on the planet and contribute to greener school grounds. They are valuable not only practically for sustainability but also for enhancing children's understanding of it. The case studies included in this chapter are truly inspirational. Adoption of ideas in this chapter will, of course, necessarily depend on your school's particular circumstances, but we hope they will inspire you, your colleagues and your pupils, sparking significant steps towards more sustainable school grounds in your setting. Taken together, small actions add up.

Key messages

- Know your school grounds and re-imagine their possibilities for nature.

- Involve your whole community from the outset to create ownership.

- Draw on expertise through national and local organisations to use resources wisely.

- Follow through shared plans and transform your playground, pupils and planet.

- Remember every action counts!

The following further resources are included on the Bloomsbury Education website:

- Examples of school ground features (photos) for the following case studies: 'Strengthening ecosystem services in Swedish preschool and school grounds' and 'Nature Friendly Schools: Stockwell Academy, Hull'.

Chapter 4
Progressions for sustainability and climate change education in lower Key Stage 2 (pupils aged 7-9)
Incorporating a cross-curricular approach to encourage the development of knowledge and skills through enquiry-based learning.

Embedding sustainability at lower Key Stage 2

This unit gives teachers and children the opportunity to develop their knowledge and understanding of the key principles of climate change and sustainability. The progressions link to the National Curriculum for England (Department for Education, 2013b) Year 3 and 4 Science and English and Key Stage 2 Geography. There is also a focus on the core principles of the Department for Education Sustainability and Climate Change Strategy (2022). In the implementation of these progressions, teachers will be encouraged to research the grounds at their school alongside their children to gain an understanding of the biodiversity in their local area. They will be required to record scientifically as they engage in a citizen scientist project. This will enable them to understand the importance of collecting data and develop their understanding of what is happening over a period and how this supports scientists nationally and globally.

Children will be encouraged to raise questions and challenge each other and their school on issues relating to sustainability, such as recycling and rewilding. They will be given the opportunity to develop an area of their school site to increase biodiversity. As such, it is important for teachers to ensure that children have secure scientific knowledge of plants, flowering life cycles and the importance of habitats and how changes can affect both animals and plants.

There are areas of this unit that can be repeated throughout the year to develop children's scientific knowledge and understanding. For example, taking part in termly citizen scientist projects will ensure children continue to develop their scientific enquiry skills and understand the importance of patterns of change over time.

Within these progressions, there are opportunities for children to work individually, in pairs or as part of a group. There are direct teaching opportunities where teacher modelling will be required, as well as more open-ended tasks.

Offering children opportunities to explore sustainability is vitally important in a world that is changing rapidly. This unit provides teachers with an exciting way of exploring this area of the curriculum with children, as well as giving them a sense of hope in a world where 'no one is too small to make a difference' (Thunberg, 2019). Giving children the chance to explore nature and raising awareness of sustainability is of great significance. If children are aware of the challenges they face and why they need to protect Earth, living in a sustainable way will become a part of everyday life for them as they preserve Earth's resources for future generations.

Natural connections

- The opportunity to explore nature
- Raising the awareness of sustainability.

Word bank for lower Key Stage 2

21st-century nature literacy

- critical thinking
- communication skills
- collaboration
- observation skills
- knowledge application
- problem-solving
- resilience
- perseverance
- creativity
- curiosity
- questioning
- planning
- organising information
- interpretation
- considering the opinions of others
- learning
- applying new skills

Technical vocabulary

- biodiversity
- protect
- planet Earth
- mapping
- research
- observation
- environment
- identification
- ecosystem
- scientific enquiry
- classification
- observable features
- names and classification of plants and animals
- citizen scientists
- observation
- data
- research
- record
- recycle
- rewild
- rethink
- review
- reflect

Summary overview

Progression	Curriculum objectives	Learning experiences/activities
Lesson 1	• KS2 Geography: Geographical skills and fieldwork • Year 4 Science: Environments and habitats	**Research:** Children will research the biodiversity of their school grounds; they will observe and identify what is in their immediate environment and map sites of interest on their school site.
Lesson 2	• Year 3 and Year 4 Science: Scientific enquiry and observation	**Record:** Pupils will become citizen scientists, collecting and recording live data from their school site. Children will submit their data to support a national survey.
Lesson 3	• Year 3 and Year 4 Science: Scientific enquiry and observation	**Recycle:** Pupils will gain an understanding of how to reuse 'pesky plastics'. They will make a water-saving device in the form of a 'tippy tap', working collaboratively as a team.
Lesson 4	• Year 3 Science: Life cycles and flowering plants • Year 4 Science: Living things and their habitats	**Rewild:** Rewilding will be the focus of this progression as children are challenged to increase the biodiversity of an area of the school site. They will make 'wild seed balls' and rewild an unused area. Children will also make their own pledge tree.
Lesson 5	• Year 3 Science: Life cycles and flowering plants • Year 4 Science: Living things and their habitats	**Rethink or take action:** Children will have the opportunity to rethink what they have learnt to date and how they can increase biodiversity on their school site. They will take part in a web of life game to support their understanding of how animals and plants rely on each other to survive. They will also make their own 'animal adaptation' from clay.
Lesson 6	• Year 3 and Year 4 Science: Scientific enquiry and observation • Year 3 and Year 4 English: Writing for different purposes	**Review, reflect and react:** In this final progression, children will reflect on the progress they have made to date. They will work as a group to come up with an action plan in an area they would like to develop further at their school, for example, a recycling campaign. They will also plant a 'seed for hope'.

Progression 1: Research
How biodiverse is the animal and plant life within the school grounds?

PREPARATION

- Source a map of the school and its grounds.
- Take 12 photos of the school site for the photo orienteering and wild mapping activities.
- Place an alphabet letter at each photo location.

Resources

- Link to video to launch unit (World Wildlife Fund, 2017)
- Inflatable planet Earth or similar
- Blindfolds
- Fact cards about planet Earth
- A range of natural objects for picture-making
- 12 photos of the school grounds for the photo orienteering and wild mapping activities
- Alphabet letters printed or drawn on paper/card
- A4 maps of the school and grounds
- Clipboards, pencils, paper and coloured dots

Previous learning

The children will have previously taken part in a photo orienteering activity and used basic maps of their school site. They will have taken part in some fieldwork to observe, measure and record physical and human features of their school site. The children will understand habitats and will be able to identify some plants and animals in their habitats.

LESSON OBJECTIVES

We are investigating the range of plant and animal biodiversity on the school site.

We are taking part in fieldwork to observe, measure and record human and physical features on a map.

We are learning the key principles of climate change and how we can take action to reduce the effects of climate change.

National Curriculum content

- KS2 Geography: Use fieldwork to observe, measure and record the human and physical features in the local area using a range of methods including sketch maps.
- Year 4 Science: Recognise that environments can change and that this can pose dangers to living things.

CONSIDER

Health and safety

Assess and evaluate hazards and risks within your setting, including any harmful plants, water sources and hazardous boundaries. Consider weather and allergies too.

Equal opportunities or differentiation

Consider how changing or adapting the space, equipment or adult support can benefit or add challenge to the activities. For example, some children may benefit from using a larger map.

INTRODUCTION

Explain to the children that they are going to discover how they can help protect planet Earth in this session. They will be taking part in activities to develop their knowledge about climate change, protecting Earth, recycling, rewilding and taking action at home and school.

Start this unit by showing the short video 'Protect our Earth' (World Wildlife Fund, 2017) or similar to illustrate to the children the challenges we face here on Earth. Make sure to emphasise that there is hope by discussing the positives in the video. Ask plenty of follow-up questions after watching: How did this make them feel? What do they want to protect on Earth? What is important to them? What did they not like about the video? What changes would they like to see made? What can they do at home and school?

WARM-UP ACTIVITY 1 (WHOLE CLASS)

Pass the globe

• Ask the children to get into a large circle and pass an inflatable globe around the circle.

• Ask key questions about our Earth. How many continents are there? How many oceans are there? Can they find any of the oceans? Can they find any of the rainforests?

• Enable them to ask a friend for help, using think-pair-share if they are unsure of the answer.

• Assess the knowledge of the children about planet Earth.

WARM-UP ACTIVITY 2 (IN PAIRS)

Human camera game

• Pair up the children and select one in each pair to be blindfolded.

• The non-blindfolded child leads their blindfolded partner to an area of the school site where they like the view. They ask their partner to lift their blindfold and use their 'human camera' to take in the view and remember the image. This is a mental snapshot taken by the children, which they will then need to recall from memory.

• The children then return to their starting point, and the blindfolded child describes the view to their partner and identifies where the view is located within the school site.

• They then swap over so both children take a 'picture'.

MAIN ACTIVITIES

Why is our Earth so important? What can we do to protect it? Encourage the children to look at the Earth fact cards to develop further understanding of what is happening on planet Earth. Ensure cards have positive facts as well as challenging facts, for example, 'the UK needs to plant 120 million trees per year by 2025' (Turns, 2022).

Challenge 1 (individual)

Earth scavenger hunt

Task the children to collect natural objects and recreate a picture of something they would like to protect. For example, make a frame using sticks and use leaves to create a picture of trees or pebbles to create a picture of animals.

Challenge 2 (in pairs)

Photo orienteering

Look at the small photos of places around the school site and ask the following questions:

• Where were these photos taken?

• Why are these sites important to the school?

• Can you find the places where the photos were taken?

• What can you find at each site?

• Can you record what you see?

Before the activity, place an alphabet letter at each photo site. Task the children to write it down next to the relevant photo. Ask the children what word(s) they can make with the 12 letters they collected. (Suggestion: 'Protect Earth').

Challenge 3 (in pairs)

Wild mapping

Look again at the small photos of places around the school site. Give the children a sketch map or digital map of the school site.

• Can they identify on the map where the photos were taken and mark these on the map with coloured dots?

• How could they make changes in those areas to support biodiversity? E.g. recycling. Can they discuss and write a sentence about each point?

Explain what biodiversity is. Example definition: 'All the different kinds of plants, animals, fungi, and microorganisms that make up our natural world' (Hancock, n.d.).

Discuss these terms with the children and model ideas, for example, 'Photo 1 is of a tap. How could we save water? Could we create something to save water in and use it to water plants?' or 'Photo 2 was a log pile. Could we create a bug hotel with the logs?' or 'Photo 3 was an overgrown area. Could we rewild this area to increase biodiversity?'

What do we need to protect on our school site? How biodiverse is our school site? What future action do we need to take? How can we make the school site more sustainable? What is sustainability? Example definition: Making different choices so that resources last longer and we can protect the planet.

EVALUATION/FOLLOW ON

What went well in the session? What knowledge and understanding do the children have about planet Earth? Where are the gaps? What do you need to do in the next session to develop further knowledge? What areas do you need to revisit in a different way?

Back in the classroom

- Create a sketch map of the school grounds and create a colour code system for the changes children could make to their school site to promote biodiversity and sustainability.

- Set up a classroom display with a map of the site with the actions children will take to build sustainability within the school.

Progression 2: Record
What is a citizen scientist and how can we become one?

PREPARATION

- Prepare some data sheets to take part in a citizen scientist project applicable to the time of year when you implement this progression. For autumn we suggest using the Bristol Climate Hub (n.d.), for spring use the RSPB (n.d.) and for summer use the Butterfly Conservation (n.d.).

- Prepare some chopped apples and cheese in advance for the session for the bird feeders if applicable to the scientific investigation.

Resources

- 10 'foreign objects' i.e. objects which would not normally be found in nature like pens or hats

- Wire, wire cutters, cheese and apples if applicable

- Recording sheets applicable to the season

- Logs, bamboo canes, clay and/or compost to make creepy-crawly towers

- Pictures of creepy-crawly towers

Previous learning

Children will have collected data and interpreted simple bar charts and graphs. Children will have taken part in an investigation and used simple recording sheets to record their observations.

LESSON OBJECTIVES

To understand what is meant by the term 'citizen scientist'.

To take part in a 'citizen scientist' investigation using scientific enquiry skills.

To understand the importance of habitats in increasing biodiversity.

National Curriculum content

Year 3 and 4 Science: Asking relevant questions using different types of scientific enquiries to answer them. Setting up practical enquiries, comparative, and fair tests. Making systematic and careful observations and where appropriate, taking accurate measurements using standard units.

CONSIDER

Health and safety

Assess and evaluate hazards and risks within your setting, including any harmful plants, water sources and hazardous boundaries. Mark or identify areas where the children may not be able to go when carrying out their investigation. Consider weather and allergies too.

Equal opportunities or differentiation

Consider how changing or adapting the space, equipment or adult support can benefit or add challenge to the activities. For example, some children may benefit from some support in recording the number of birds they see, in which case, provide them with an example recording sheet.

INTRODUCTION

Explain to the children what it means to be a 'citizen scientist'. Explain how important the role is in supporting scientists across the globe by learning more about local environments, identifying patterns of change and noticing the decline or increase in species numbers. Identify a national citizen scientist survey that the children could take part in (ensuring the survey is relevant to the time of the implementation of this unit).

Warm-up activity (whole class)

The super-power observers

Talk to the children about 'observation'. Observation and taking notice of what is happening in your local environment are very important skills as citizen scientists. Explain to the children that they are going to form a line and follow you on an observation trail. They must be very quiet so as not to disturb any wildlife. On the trail, they will need to individually look for at least 10 'foreign objects' which have been placed along the trail. These objects are not hidden but may be hard to see if you are not looking closely. Explain that foreign objects are objects which would not normally be seen in that area e.g. a pen in a tree.

How many objects did they see? Which objects were difficult to identify at first glance? What skills of observation did they use to find the objects?

MAIN ACTIVITIES

Explain to the children that they will now take part in some challenges to prepare for a citizen scientist investigation. Remind them of the skills they have been practising in the initial activities. What skills are important to being a successful citizen scientist? How can we develop these skills?

Challenge 1 (individual)

Fruit loops

If the children are going to take part in the 'Big Bird Watch', help them to make some fruit loops before taking part in the survey. These can be made by giving each child a length of garden wire and threading chunks of cheese, apples, and raisins onto the wire. Close the wire when the threading is complete and then tie them up with string where the 'Big Bird Watch' will take place to attract birds to feed.

Challenge 2 (individual)

Citizen scientist

Children will now take part in a citizen scientist investigation. Remind the children what a citizen scientist is and how important they are to understanding the impacts of climate change locally, globally and nationally.

Give them the relevant recording sheet and point out key features that will aid identification. Allow the children time to make their observations. Encourage them to find a quiet place where they can sit and take note of the birds (or other survey-specific wildlife) they can see. Discuss what the children have found. Are there common patterns and themes in the whole class data? What was challenging about watching and identifying? How might this data now be used to inform future studies?

Challenge 3 (pairs or small groups)

Create a creepy crawly tower

Talk to the children about what habitats are in the school. Remind the children what biodiversity is, i.e. the number of different plants and animal species living in an area. What animals live in the school's outdoor space? What sorts of habitats do insects and other minibeasts need? Can the children increase biodiversity by making habitats for animals?

Provide the children with some basic equipment to build their own creepy crawly tower. Children can work in pairs or small groups of three or four. Look at the images of some creepy crawly towers made with bricks, wood, stones, mud and small pallets for inspiration. Give the children time to build their tower.

Once built, ask the children if they can predict what species may use their tower. Why would these creatures like to live there? What may they see when they return in one to two weeks? How could they record the species that use the tower? How could this support a citizen scientist project?

PLENARY

Talk to the children about what they have observed. What have they recorded on their observation sheets? Plan for future investigations building on their scientific investigation skills. Ensure the children can input their data onto a live investigation if relevant so they further develop their understanding of the importance of taking part in citizen scientist investigations.

EVALUATION/FOLLOW ON

What went well in the session? What strengths do the children have in carrying out investigations? What areas do they need to improve? What further support do they require? What areas of knowledge need to be revisited in the next session?

Back in the classroom

- Encourage the children to design their own unique bug hotels, creepy crawly towers or other animal habitats. What other materials can they use? Could this be linked to a design technology project? Discuss the importance of habitats and the link to biodiversity in reducing the effects of climate change. Link to the use of natural and recycled materials to make these manmade ones sustainable.

- Look at the data the children have collected. Can this be linked to a maths project? Can they develop line graphs or bar charts to represent the data collected?

Progression 3: Recycle
What is recycling and how can it support us in reducing the effects of climate change and promote sustainability?

PREPARATION

Prepare some question and answer cards in advance. Questions could be printed on one colour card with answers on another colour card. These should relate to previous learning. Some example questions include: What is sustainability? What do we mean by rewilding? What is biodiversity? See Appendix 8 for further examples of questions and answers.

Resources

- Blindfolds
- Recycling fact cards
- Question and answer cards
- Different coloured, labelled hoops
- Items (or pictures of items) that can be recycled, reused, repaired or reduced
- Audit sheets for lunch waste audit
- Empty plastic milk bottle with lid, string, nails, matches and tea lights

Previous learning

Children will have a basic understanding of which items can be recycled. They will have collected data and completed basic recording charts before.

LESSON OBJECTIVES

To understand the key principles of recycling, repairing, reusing and reducing.

To complete a lunch waste audit, take action as a class and work towards reducing lunch waste.

To make a water-saving device which could be used in the outdoor learning area.

National Curriculum content

- Year 3 and 4 Science: Asking relevant questions using different types of scientific enquiries to answer them. Setting up practical enquiries, comparative, and fair tests. Making systematic and careful observations and where appropriate, taking accurate measurements using standard units.

CONSIDER

Health and safety

Assess and evaluate hazards and risks within your setting, including any harmful plants, water sources and hazardous boundaries. Consider weather and allergies too.

Equal opportunities or differentiation

Consider how changing or adapting the space, equipment or adult support can benefit or add challenge to the activities.

INTRODUCTION

Talk to the children about the need for us to recycle, reduce, reuse and repair. To tackle climate change, we need to reduce the amount of rubbish we send to landfill sites. Discuss some recycling facts. Discuss the idea that we need to work as a team at school, in the community and at home if we are going to take action and make a difference.

Warm-up activity 1 (groups of three)

Recycling robot game

Talk to the children about the need to be excellent at communication and teamwork. Introduce the children to the recycling robot game to improve this. See Figure 4.1.

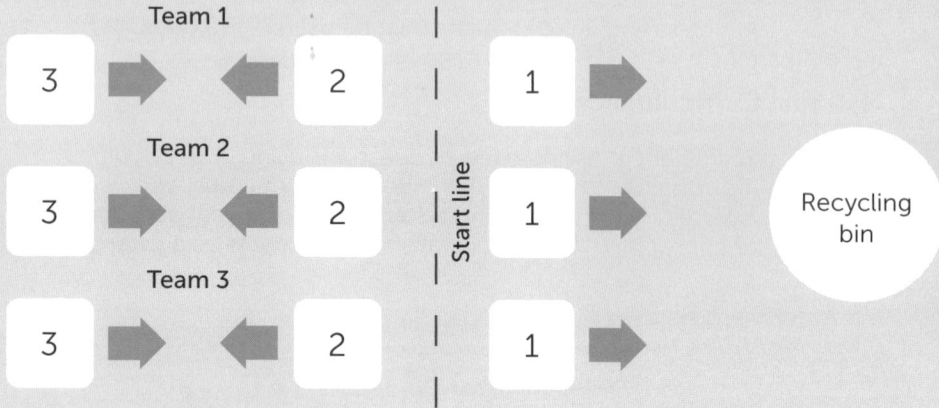

Figure 4.1: Recycling robot game.

Child 1 is the Robot and is blindfolded, facing the recycled bin and is the only participant permitted in the activity area.

Child 2 is the Communicator and faces away from the activity area – they must NEVER observe the Robot.

Child 3 is the Observer and faces the activity area, they cannot talk at all and can only communicate non-verbally with the Communicator.

Can the children communicate with the Robot to collect the recycled material? The objective is for robots to retrieve the recycled item from the activity area before the other teams. This can be played in teams of three collecting one object at a time.

Warm-up activity 2 (small groups)

Hoops relay game

Working in teams, children sort the items or pictures of items into reuse, recycle, repair or reduce using four different coloured labelled hoops. When all the items have been sorted, discuss with the children why they chose those hoops. Do any of the items fit into more than one category? Are any items not in the correct hoop?

MAIN ACTIVITIES

Complete a paired question activity to recap on previous learning using the question and answer cards. The children must then find the matching question and answer. This will support the building and development of children's knowledge of key concepts.

Challenge 1 (whole class)

Lunch waste audit

Ask the children to look into their lunchboxes. Ask them questions like: Can we reduce the waste we produce at lunchtimes? Which items could be recycled? How can we recycle them? Conduct an audit of lunch waste for the whole class and use a simple recording chart to record different items of waste. As a whole class, look at the data. Which is the highest amount of waste produced? How can we reduce the amount of waste we produce at lunch? How could we increase the amount of recycling and composting we do as a school?

Challenge 2 (small groups)

Pesky plastics

Look at the range of plastic items. Working in groups, sort the items into two categories: reuse and recycle. Ask the children to come up with some ideas as to how the plastic items could be reused. Can they be used to improve the outside space? Can they be reused to support sustainability? Give the children time to explore and discuss how the plastic items that have been collected could be reused.

Extension activity (small groups)

The tippy tap

Task the children with making a tap for the outside space using an empty milk bottle. They will need string, nails and some matches. This is an idea being developed in many African countries, for more information, visit www.tippytap.org. Discuss how useful the tippy tap is in developing countries, but also in any outdoor area where there may not be running water or as a way of reducing water use.

PLENARY

Discuss with the children their understanding of the importance of recycling, reducing, reusing and repairing. Talk to the children about how they can take action within the school, referring to the map they developed of the school site previously. Where are the recycling points in the school? What else could they recycle at school? Can they use the compost bin more effectively? How can they reduce water use?

EVALUATION/FOLLOW ON

What went well in the session? What strengths do the children have in carrying out investigations? What areas do they need to improve? What further support do they require? What areas of knowledge need to be revisited in the next session?

Back in the classroom

- Investigate the tippy tap concept and look at the website to understand the concept further (www.tippytap.org).

- Look at other water-saving devices that the school could use. Can the children design their own water-saving device? Could they develop a recycling campaign to increase the number of items recycled? Could they share their findings with the whole school and parents in an assembly?

PREPARATION

Before the session, identify an area of the school site that you could rewild. Ensure it is safe and accessible for the children. Tools and suitable gloves may be needed to support the children in clearing the area.

Resources

- Blindfold
- 2 parts compost, 5 parts clay, 1 part water, 1-2 parts wildflower seeds and a mixing tub
- Garden tools and gloves
- Old willow branches or similar
- Wooden discs or leaves
- String

Previous learning

Children will have a basic understanding of how seeds and plants grow into mature plants.

Children will be able to describe how plants need water, light and a suitable temperature to grow and stay healthy.

LESSON OBJECTIVES

To develop an understanding of what is meant by the term 'rewilding'.

To develop knowledge and understanding of methods that can be used to rewild an area of the school.

To learn how to make a wild seed ball.

National Curriculum content

- Year 3 Science: Explore the life cycle of flowering plants, including pollination, seed formation and seed dispersal.
- Year 4 Science: Explore and use classification keys to help group, identify and name a variety of living things in the local and wider environment.

CONSIDER

Health and safety

Assess and evaluate hazards and risks within your setting, including any harmful plants, water sources and hazardous boundaries. Consider weather and allergies too.

Equal opportunities or differentiation

Consider how changing or adapting the space, equipment or adult support can benefit or add challenge to the activities.

INTRODUCTION

Ask the children questions like: What plant species live in the local area? How important are trees to reducing climate change? Why do we need trees? Discuss with the children the importance of trees in producing oxygen, storing carbon, stabilising the soil and providing habitat/food for wildlife. Explain that trees also provide shelter and the materials for tools.

Warm-up activity (pairs)

Hug a tree

In pairs, one child is blindfolded and the other is the nominated guide. Ensure the children are given clear boundaries to work within and ensure they always guide their partner safely. Explain to the children that they need to lead their blindfolded partner to a tree. If there are no trees in the area, ask the children to find an interesting natural object. On arrival, they need to touch the tree and 'hug' it if possible. The guide then needs to lead their partner back to the starting position. The blindfolded child must now remove their blindfold and guess which tree they were taken to. What did they notice about 'their' tree? What characteristics made it memorable? Encourage the children to tie a ribbon on their favourite tree and discuss why we need to protect our trees.

Ask the children some conversation-starting questions: How old is your favourite tree? Measure it using hand spans (the width of your hands). How many hand spans can you fit around your favourite tree? Your hand span represents approximately 5 years, so how old is your favourite tree? How tall is your tree? Stand with your back to the tree and take some large strides, approximately one metre long. Count the number of one-metre strides you take. Keep moving away from your tree until you can see the top of the tree through your legs. How many strides did you take? How tall is your tree?

MAIN ACTIVITIES

Talk to the children about the concept of rewilding. Discuss potential areas on the school site where rewilding could take place. Do they have an area at home that they could rewild? Remind the children why rewilding is important and the impact this can have in creating a sustainable environment.

Challenge 1 (individual)

Wild seed ball

Discuss wildflowers and how they support animal species, pointing out their particular importance for pollinating species. Make a seed ball to rewild an area of your school. You will need: 2 parts compost, 5 parts clay, 1 part water, 1-2 parts wildflower seeds and a mixing tub.

1. Mix the soil, clay and water thoroughly to make balls. Add the seeds and keep mixing until the seeds are within the balls. The balls should hold together easily. They should be the size of a small golf ball. You will need to dry the seed balls for a few days, ideally in the sunshine.

2. Scatter the seed balls onto bare, undug soil. The wildflower seeds are wrapped up in clay and compost – the outer layer of the clay helps to prevent birds and insects from eating the seeds.

3. Seeds grow inside the ball and begin to sprout. With rain (or watering) and warmth, the wildflower seeds inside the balls will start to sprout. This can take up to 4-6 weeks from the time you scatter the balls.

4. The balls crumble as the plants grow, leaving a patch of young plants which will continue to grow and grow into a wildflower garden!

Challenge 2 (small groups or whole class)

Tree naming

Plant a tree or plant and have a celebratory naming ceremony. Take time to prepare some ground within your school to plant this tree or plant.

Challenge 3 (individual)

Pledge tree

Make a pledge tree using an old willow branch. Encourage the children to pledge how they are going to live more sustainably and reduce the effects of climate change. Write each pledge on a wooden leaf or disc and hang it on the pledge tree. Use this as a display to promote your work across the school.

Talk to the children about the activities they have taken part in during this progression and the importance of trees and wildflowers for biodiversity. What have they learned today about trees and wildflowers? How has this supported them in their understanding of biodiversity on the school site?

EVALUATION/FOLLOW ON

What went well in the session? What areas do they need to improve? What further support do they require? What areas of knowledge need to be revisited in the next session?

Back in the classroom

- Investigate the importance of trees by carrying out a survey of trees across the school site or local park.
- Write letters to the Woodland Trust to see if the school can access some free trees.

Progression 5: Rethink or take action
What we have learned so far and how can we take action to increase biodiversity on our school site?

PREPARATION

- Prepare picture cards that outline the journey of a seed.
- Prepare cards with the names of animals, plants and organisms on them (including 'tree').

Resources

- Mirrors
- Picture cards (laminated A4) that outline the journey of seed to flower back to seed
- Cards with the names of animals, plants and organisms on them
- Ball of string
- Tubular webbing tape
- Clay, pebbles, stones or other natural items for children to create their own animals
- Popcorn kernels, oil, a pan and an outdoor stove
- Old newspaper and plastic bottles

Previous learning

Children will have an understanding of habitats and will be able to describe how different habitats provide for the basic needs of different types of animals and plants, and how they depend on each other.

LESSON OBJECTIVES

To understand how an ecosystem works.

To understand how plants and animals depend on each other within an ecosystem.

National Curriculum content

- Year 3 Science: Explore the life cycle of flowering plants, including pollination, seed formation and seed dispersal.
- Year 4 Science: Recognise that environments can change and that this can sometimes pose dangers to living things.

CONSIDER

Health and safety

Assess and evaluate hazards and risks within your setting, including any harmful plants, water sources and hazardous boundaries. Consider weather and allergies too.

Equal opportunities or differentiation

Consider how changing or adapting the space, equipment or adult support can benefit or add challenge to the activities.

INTRODUCTION

Show the children the 'seed to plant' video (Ms. Lindsey's Book Nook, 2021) or use a similar resource which shows the story of seed to plant. Talk about the wild seed balls made in the previous session. What has happened to the seed balls? Are they beginning to grow roots and shoots? Reflect on the outside space and consider what still needs to be done. Revisit the creepy crawly towers or bug hotels and see if any creatures have moved in.

Use these as a lens through which to study ecosystems and the 'web of life' to understand the importance of animals and plants in supporting each other.

Warm-up activity 1 (pairs)

Squirrel climb

Encourage the children to take another look at who lives in their outside space. In pairs, ask them to take a mirror and place it on the bridge of their nose. Encourage the children to choose a tree they would like to look at more closely. The children should start at the base of the tree with the mirror on the bridge of their nose and 'climb' slowly up the tree by moving the mirror image up the tree. As the children move the mirror up the base of the tree, they are tilting the mirror on their nose to effectively walk up the tree. What do they notice?

Warm-up activity 2 (whole class)

Circle questions

Recap previous learning points as a group standing in a circle made from tubular webbing tape. Ask the children a question (e.g. what is biodiversity?) and ask them to step into the circle if they know the answer. Alternatively, ask the children to stand in a circle around the tape. When you say 'stop' encourage the children to tell you something they have learned or ask a question they would like to know more about.

MAIN ACTIVITIES

Discuss biodiversity with the children. What does it mean? How can planting flowers and plants within an area contribute to biodiversity? Why is biodiversity important to sustainability within our environment?

Challenge 1 (pairs)

Life cycle of a flowering plant

Talk to the children about the story of a seed. Present them with picture cards that outline the journey of seed to flower back to seed. Ask the children to look at the cards and see if they can correctly order the pictures. Write a number on each picture to order them.

Challenge 2 (whole class)

The web of life game

For this game, use the pre-prepared cards with the names of animals, plants and organisms on them and a ball of string. Ask the children to sit in a circle. Place the cards in the middle of the circle. Each child takes a card from the pile in the middle and holds it so everyone can see. The child with the 'tree' starts with a ball of string, holds onto one end, and throws it to someone else in the circle. The person who catches the string explains how the organism on their card interacts with the tree. They then hold onto the string before throwing it to a third person. The game continues until everyone in the group is interconnected by string. Choose one of the organisms and ask the children to predict what would happen if they were removed. What other organisms would be affected? Discuss how plants and animals depend on each other to survive in an ecosystem.

Challenge 3 (individual)

Animal adaptations

Give the children the opportunity to create their own unique model of an animal. Using clay and natural items, let them create their own unique animal. They will need to give it a name, and consider where would it live and what it would eat. Also, consider how it would adapt to its environment and cope with climate change.

Challenge 4 (whole class)

Extension

Cook some natural popcorn in a pan with some oil in the outdoor area. Talk to the children about how pieces of popcorn are in fact 'exploded seeds' and about the plant they come from. Can they name any other forms of corn we eat? Discuss other seeds, like wheat used to make the flour for bread. Recap how seeds are formed and how they grow, and discuss the life cycle of a seed and flowering plant.

Talk to the children about ecosystems – how plants and animals depend on each other to survive and how they adapt to survive. Discuss why it is important to protect ecosystems and increase biodiversity.

What went well in the session? What strengths do the children have in carrying out investigations? What areas do they need to improve? What further support do they require? What areas of knowledge need to be revisited in the next session?

Back in the classroom

- Make plant pots from old newspapers. Wrap strips of waste paper tightly around a bottle several times before folding over the base and sliding the paper off the bottle. Fill the pot with compost and plant a seed. Look after the plant at home before bringing it back into school to plant.

A classroom display detailing some of this class's completed progression work.

PREPARATION

Before the session, prepare laminated cards with examples of how children have taken action at their school (see Appendix 9 for suggestions).

Resources

- Wooden building blocks with the names of different organisms written on them

- Five hoops labelled: recycling, rewilding, reducing, re-using, climate change

- Examples of when children have taken action printed on laminated card

- Plant pots, compost, and seeds

Previous learning

In this final session, children will utilise the knowledge they have acquired in the previous sessions to develop a plan of action in an area they are interested in. For example, they could develop a plan to raise some money to plant a wildflower garden.

LESSON OBJECTIVES

To understand how we can take action within our school to reduce the effects of climate change.

To revisit the importance of ecosystems in 'the tree of life'.

To develop a sustainable plan of action for the future.

National Curriculum content

- Year 3 and Year 4 Science: Asking relevant questions using different types of scientific enquiries to answer them. Setting up practical enquiries, comparative, and fair tests. Making systematic and careful observations and where appropriate, taking accurate measurements using standard units.

- Year 3 and 4 English: Pupils should have opportunities to write for a range of real purposes and audiences as part of their work across the curriculum.

CONSIDER

Health and safety

Assess and evaluate hazards and risks within your setting, including any harmful plants, water sources and hazardous boundaries. Consider weather and allergies too.

Equal opportunities or differentiation

Consider how changing or adapting the space, equipment or adult support can benefit or add challenge to the activities.

INTRODUCTION

Talk to the children about the areas of the curriculum covered to date. In this progression, explain that they will be investigating how to take further action at the school. Children will be working in small groups to develop an action plan. This will require them to communicate, work as a team and collaborate.

MAIN ACTIVITIES

Tell the children that they will take part in two different team-building activities to develop their team-building and collaboration skills.

Challenge 1 (whole class)

Wooden block tower

Ask the children to build a tower using wooden blocks with the names of different organisms written on them. Explain to the children that the tower is an ecosystem representing a tree of life. The organisms in the tower rely on each other to survive. Ask the children to take turns removing a single block. What happens to the tower? How long does it stay standing? Much like the ecosystem, each organism depends on each other for food and life. If one organism is removed, there is an 'imbalance' in the ecosystem and plants and animals will struggle to survive.

Challenge 2 (small groups)

Taking action

Talk to the children about the need to take action to ensure sustainability at the school. Using five hoops labelled recycling, rewilding, reducing, reusing and climate change, place the children in groups and give them a set of scenario cards. Some examples of these cards are listed in Appendix 9. Each group needs to decide which hoop the card fits into.

Challenge 3 (individual)

Plant a seed for hope

Using compost, seeds and a recycled plant pot made in a previous session, give the children a seed to plant for 'hope' (which they can keep in the class or take home). Explain to the children that the seed is going to take time to grow and will need to be cared for to ensure it flourishes. The same can be said for developing a sustainable school – their action plans will need to be carefully thought out if they are going to make a real difference. However, if they are fully committed, they can make changes towards a sustainable future.

PLENARY

Discuss with the children the activities of the past six progressions. Which topic areas did they enjoy the most? What areas would they like to learn more about? How are they now going to take action?

EVALUATION/FOLLOW ON

What went well in the session? What strengths do the children have in carrying out investigations? What areas do they need to improve? What further support do they require? What areas of knowledge need to be revisited in the next term? What topic areas did the children enjoy? What are the next steps in developing their knowledge and learning about sustainability and climate change?

Back in the classroom

- Provide the children (working in the same small groups as challenge 2) with a template action plan with the following headings: What are we going to do? How are we going to do it? What resources will we need? What difference will it make to sustainability at our school?

Chapter 5
Progressions for sustainability and climate change education in upper Key Stage 2 (pupils aged 9-11)
Incorporating rich cross-curricular links embedding knowledge and skills through active and authentic approaches

Embedding sustainability at upper Key Stage 2

This unit offers the opportunity for teachers and educators to develop some of the knowledge and skills needed to empower children to restore nature within the context of sustainability and hope. The following six progressions provide the opportunity to use outdoor areas around the school to learn more about nature and climate change from a local perspective. They are based on the National Curriculum for England (2013b) Year 5 and Year 6 Science, English and Geography content and the Department for Education Sustainability and Climate Change Strategy (2022).

By spending more time in and with nature, these units aim to raise a greater awareness of how climate change is affecting animal and plant phenology (the study of how the biological world times natural events, usually through seasonal changes). In these progressions, children aged 9 to 11 work scientifically, using and applying knowledge from classroom learning to real-life investigations. Regular, repeated species recording provides useful data, which can be gathered and analysed to provide insights into how a setting changes over time. Repeating this unit over the academic year(s) provides the opportunity to observe and compare seasonal life-cycle changes in a variety of living things in the local environment. It also provides the chance to study animal and plant phenology in a longer-term citizen science project into the effects of climate change. Within this context, the children raise questions about the diversity of their local environment, recording data, critically reflecting and reporting their results. They use the results of their investigations to develop specific areas of the school grounds to encourage growth and biodiversity in the outdoor learning area. This is part of a longer-term project of hope and positivity.

The rapid changes in our world require educators to adopt a flexible approach to teaching and learning, and so these progressions provide opportunities for critical thinking, communication, creative thinking and collaboration, as well as the development of technology literacy. There are opportunities for the children to work independently and in small groups or pairs (with adults providing positive support), in addition to whole-class teacher-directed learning.

Key impact indicators

- Observation of nature
- Raised awareness and developing knowledge of conditions required to thrive and survive, plant adaptations, stewardship roles and the impact of humans on nature
- Scientific study and classification of living things
- Interpretation of information, building core values and attitudes for sustainability, developing a nature connection, compassion and empathy for nature.

Word bank for upper Key Stage 2

21st-century nature literacy

- critical thinking
- communication skills
- collaboration
- observation skills
- knowledge application
- maintaining concentration
- problem-solving
- resilience
- perseverance
- self-direction

- creativity
- curiosity
- questioning
- planning
- organising
- information
- interpretation
- justifying viewpoints
- considering the opinions of others
- learning and applying new skills

Technical vocabulary

- biodiversity
- phenology
- ecosystem
- scientific enquiry
- classification
- observable features
- names and classification of plants and animals

Summary overview

Progression	Curriculum objectives	Learning experiences/activities
Lesson 1	• Year 5 Science: Living things and their habitats • Year 6 Science: Living things and their habitats • UKS2 Science: Working scientifically • KS2 Geography: geographical skills and mapwork	**Research:** Children will investigate the range of animal biodiversity in their outdoor area. They will use the terms 'ecosystem' and 'biodiversity', and deepen their understanding of the importance of biodiversity in responding to climate change. They will do a setting baseline survey, recording evidence of animals and their habitats.
Lesson 2	• Year 5 Science: Living things and their habitats • Year 6 Science: Living things and their habitats • UKS2 Science: Working scientifically • KS2 Geography: geographical skills and mapwork	**Record:** Pupils will investigate the range of plant biodiversity of the outdoor area by planning and carrying out a setting baseline survey, recording evidence of plants.
Lesson 3	• Year 5 Science: Living things and their habitats • Year 6 Science: Living things and their habitats • UKS2 Science: Working scientifically • KS2 Geography: geographical skills and mapwork	**Roles:** Children investigate why some areas have greater biodiversity than others. They will discuss the importance and role of biodiversity in maintaining ecosystems and combatting climate change, gathering information to explain, raise questions or hypothesize about the impact of differences in biodiversity.
Lesson 4	• Year 5 Science: Living things and their habitats • Year 6 Science: Living things and their habitats • UKS2 Science: Working scientifically • KS2 Geography: geographical skills and mapwork	**Revisit:** Pupils will learn about the interactions, connections and interdependence of an ecosystem. A web game will show the importance and role of biodiversity in maintaining ecosystems. They will revise the phenology and life cycle of a flowering plant, then use orienteering to look for and collect seeds. They will consider new planting as a biodiversity factor.
Lesson 5	• Year 5 Science: Properties and changes of materials • Year 5 Science: Living things and their habitats • Year 6 Science: Living things and their habitats • Year 6 Science: Evolution and inheritance • UKS2 Science: Working scientifically	**Reproduction:** Factors that influence and promote growth will be explored in this progression. Pupils will analyse soil quality for pH, temperature and type, before experimenting with planting in different soils. They will explore reproduction in plants by planting in pots and containers as part of a micro-growing experiment.
Lesson 6	• Year 5 Science: Living things and their habitats • Year 6 Science: Living things and their habitats • UKS2 Science: Working scientifically	**Rejuvenate:** Pupils will use what they now know to increase the biodiversity and develop a micro national park on the school grounds. They will plan and take action to improve biodiversity to mitigate the effects of climate change.

Progression 1: Research
How biodiverse is the animal life of the school setting?

PREPARATION

- Create big question cards using ideas from Appendix 10.
- Create maps of your outside area identifying key features (walls, pond, playground boundaries, significant individual or groups of trees, planting areas, hedges, etc.).

Resources

- Big question cards using ideas from Appendix 10
- Clipboards, pencils, paper (or sketchbooks) and coloured pencils
- Large A3 maps of the area with key features identified
- Dichotomous keys, ID sheets or species ID apps
- Quadrats or hoops
- Sweep nets
- Pond-dipping nets
- White sheets
- Thermometers
- Magnifying glasses
- Trays
- Cameras or tablets
- Optional: Google Maps, GPS or geo-location apps

Previous learning

The children will have previously used maps to explore the outdoor area.

In Years 3 and 4, children will have defined ecology and biodiversity, and explored living things and their habitats, including invertebrate group classification.

LESSON OBJECTIVES

To investigate the range of animal biodiversity of the outdoor area.

To understand the terms phenology and biodiversity.

To understand that different areas support different biodiversity.

To understand the importance and role of biodiversity in relation to climate change.

National Curriculum content

- Year 5 Science: Study and raise questions about their local environment throughout the year, asking pertinent questions and suggesting reasons for similarities and differences. Describe the differences in the life cycles of a mammal, an amphibian, an insect and a bird. Describe the life process of reproduction in some animals.

- Year 6 Science: Describe how living things are classified into broad groups according to common, observable characteristics based on similarities and differences. Give reasons for classifying animals.

- UKS2 Science: Work scientifically by planning different types of scientific enquiries to answer questions, recording data and reporting and presenting findings.

- KS2 Geography: Use fieldwork to observe, measure and record the human and physical features of the local area using a range of methods, including sketch maps.

CONSIDER

Health and safety

Assess and evaluate hazards and risks within your setting, including any harmful plants, water sources and hazardous boundaries. Consider weather and allergies too.

Equal opportunities or differentiation

Consider how changing or adapting the space, equipment or adult support can benefit or add challenge to the activities.

INTRODUCTION

Explain to the children that they are 'citizen scientists' (this will be a familiar concept to those who have completed the LKS2 progressions). They will be investigating how the outdoor areas, climate, plants and animals interact to create areas rich in diversity (or not), and to see if this is changing over time as a result of climate change.

Ask the children: Why is biodiversity important? The more biodiversity there is, the stronger the ecosystem will be. This means small changes, like those relating to climate change, will have less of an effect. As citizen scientists, the data they collect today will be compared to data collected at specific times and dates throughout the year and in future years, providing information that will help monitor the effects of climate change.

Today, they will be conducting a baseline survey to find out the answer to the following question: 'How biodiverse is our setting in animal life?'

In small groups or pairs, children share their thoughts on the following questions with a partner before feeding back to the whole class:

1. Scientifically, how might you find out the answer to this question? (Focus on fairness and accuracy of recording, e.g. recording the date and time of day, location or weather conditions).

2. What do you think you will find?

WARM-UP ACTIVITY (PAIRS)

Think-pair-share big questions

Assess existing knowledge using big question cards (see Appendix 10 for suggestions) to think-pair-share. In pairs, children ask each other questions to find out what they know about animal classification, animal identification or potential habitats. This is excellent practice for developing subject-specific vocabulary.

MAIN ACTIVITIES

Ask children if they think there is evidence of good biodiversity in the school setting. Remind them that the key purpose of this activity is to provide information that will help monitor the effects of climate change in this two-part scientific challenge.

Challenge 1 (pairs or groups of four)

Identify and record

Children will identify and record the location of animals and their habitats in the outdoor area. First, demonstrate the survey gathering techniques (dichotomous keys, species apps), tools and scientific recording (data gathering, taking photographs). Remind the children that they will be looking for evidence of invertebrates, birds, large mammals, small mammals (e.g. rodents and bats), amphibians (e.g. frogs and toads), reptiles (e.g. snakes and lizards, including slow worms) and perhaps fishes (if there is a pond area). This will be achieved using their knowledge of life cycles to look for clues of animal life forms.

After suitable teacher-led activity modelling, children should:

- use a quadrat or hoop at specific locations on the map (control points)

- use sweep nets over grass

- go pond dipping for invertebrates

- place a white sheet under a tree and shake the tree

- bird watch over time (which can be linked to DT and maths by building a hide)

- take the temperature of the pond water

- look for tracks or droppings as evidence of larger animals and birds

- sketch or make details drawings of what is found (linking to KS2 Art)

- take macro and close-up photographs (linking to KS2 Art)

- identify plans by observable characteristic using dichotomous keys, ID sheets or species ID apps (which can be used as part of a wider citizen science data gathering task)

- if available, log their findings using tools such as Google Maps, GPS or geo-location apps.

Emphasise the need to leave habitats as they were found, discussing reasons for this. For example lift, look and lower the logs gently before replacing them in their original positions.

The groups should explore the outdoor learning area, collecting and recording specific data and using the setting map to plot the animal diversity of the grounds as shown.

Challenge 2 (groups of four)

Assess your findings

Following a thorough investigation at each location in the setting, as a group, the children agree on a Red/Amber/Green (RAG) grade for the diversity levels of that location. For example, red (little evidence of animal life), amber (a few) or green (lots), and marking these on the map. This will be used to inform future choices as part of a wider scientific investigation of the site.

PLENARY

Process the information collected. As a whole group, reflect on the numbers in the findings. Do all groups agree on the RAG rating for each area? What are their conclusions about the animal biodiversity of the setting?

To reflect on the session, pairs of children can consider what they have found out by telling another pair: 3 things they learned from doing this activity, 2 things they want to know more about and 1 question they would like the answer to.

EVALUATION/FOLLOW ON

Encourage the children to explain what they have found out about the diversity of animals in the area. Do the groups agree? Were they surprised by their findings? What were the challenges of this investigation and how were these overcome? What went well and why? What didn't go as well as expected? What could be changed? Who stood out and why? What do they now know about sustainability and the impact of climate change that they didn't know before? Who will they share this new learning with?

Back in the classroom

- Interpret, display and compare the data that has been collected. Create graphs to illustrate which habitat was the most diverse in terms of invertebrates, birds or animals.

- Using computing, write a non-chronological report of their investigation, or a 'how-to guide', for future classes doing this survey. Reinforce the purpose of the activity – to monitor and assess the impact of climate change on the setting over time.

- Plot the GPS locations on an e-map so that accurate data can be collected over time from the same location.

- Follow up on the plenary activity by finding out more about the 2 'things' and develop a plan to find out the answer to their 1 question.

- Find more about the classification of invertebrates using such terms as 'arthropod' and 'phylum euarthropoda'.

- Find out how to encourage more biodiversity in a garden (consider an action plan).

PREPARATION

- Create big question cards using ideas from Appendix 11.

- Using the master maps of your outside area from Progression 1, ensure key features are identified (walls, pond, playground boundaries, trees, planting areas, hedges etc.).

- Research plant species likely to be present in the setting, including grasses, herbaceous plants (plants with flexible, green stems and no woody parts) and woody shrubs and trees.

Resources

- Large A3 maps of the area with key features identified

- Clipboards, paper (sketchbooks), pencils and coloured pencils

- Plant species dichotomous keys, ID sheets or species ID apps

- Magnifying glasses

- Quadrats or hoops

- Cameras or use tablets

- Optional: Google Maps, GPS or geo-location apps

Previous learning

Map skills, categorisation, RAG rating and data gathering in Progression 1.

In Years 3 and 4, children will have defined ecology and biodiversity, and explored plants and plant identification.

LESSON OBJECTIVES

To investigate the range of plant biodiversity of the outdoor area.

- To understand that different areas support different biodiversity.

- To understand the importance and role of biodiversity in relation to climate change.

National Curriculum content

- Year 5 Science: Study and raise questions about their local environment throughout the year, asking pertinent questions and suggesting reasons for similarities and differences.

- Year 6 Science: Describe how living things are classified into broad groups according to common, observable characteristics based on similarities and differences.

- UKS2 Science: Work scientifically by planning different types of scientific enquiries to answer questions, recording data and reporting and presenting findings.

- KS2 Geography: Use fieldwork to observe, measure and record the human and physical features of the local area using a range of methods, including sketch maps.

CONSIDER

Health and safety

Assess and evaluate hazards and risks within your setting, including any harmful plants, water sources and hazardous boundaries. Consider weather and allergies too.

Equal opportunities or differentiation

Consider how changing or adapting the space, equipment or adult support can benefit or add challenge to the activities.

INTRODUCTION

Explain that today the children continue their journey as 'citizen scientists'. They will be investigating how the outdoor areas, climate, plants and animals interact to create areas rich in diversity (or not), and to see if this is changing over time as a result of climate change. Reinforce the purpose of the activity: to monitor and assess the impact of climate change on the setting over time. As citizen scientists, the data they collect today will be compared to data collected at specific times and dates throughout the year and in future years, providing information that will help monitor the effects of climate change.

Today, they will be conducting a baseline survey to find out the answer to the following question: 'How biodiverse is our setting in plant life?'

Using knowledge and skills practised in Progression 1, take time to reflect on:

1. What does this question mean?

2. Scientifically, how might you find out the answer to this question? (This builds on knowledge and skills gained from Progression.)

3. What do you think you will find? (This provides the opportunity to share prior research.)

Think-pair-share: big questions

Assess existing knowledge using big question cards (see Appendix 11 for suggestions) to think-pair-share. In pairs, children ask each other questions to find out what they know about plant classification and identification, and practice/develop subject-specific vocabulary.

MAIN ACTIVITIES

Ask children if they think there is evidence of good plant biodiversity in the school setting.

Remind them that the key purpose of this activity is to provide information that will help monitor the effects of climate change over time in this two-part (Progressions 1 and 2) scientific challenge.

Challenge 1 (in pairs or groups of four)

Identify and record

Children will identify and record the location of plants in the outdoor area. First, demonstrate the survey gathering techniques, tools and scientific recording. Then explain that the children will be looking for and identifying grasses, herbaceous plants (plants with flexible, green stems with no woody parts), woody shrubs and trees.

After suitable teacher-led activity modelling, children should:

- use a quadrat or hoop at specific locations on the map (control points)

- sketch or make detailed drawings of what is found (linking to KS2 Art)

- take macro and close-up photographs (linking to KS2 Art)

- identify plans by observable characteristics using dichotomous keys, ID sheets or species ID apps (which can be used as part of a wider citizen science data gathering task)

- if available, log their findings using tools such as Google Maps, GPS or geo-location apps.

Emphasise the need to leave habitats as they were found, discussing reasons for this.

The groups then explore the outdoor learning area, collecting and recording plant data, using the setting map to plot the plant diversity of the grounds at specific control points. Use think-pair-share to clarify the tasks. They will be considering the implications of this in relation to increasing diversity in the setting.

Challenge 2 (groups of four)

Assess your findings

Following a thorough investigation at each location in the setting, as a group, the children agree on a Red/Amber/Green (RAG) grade for the diversity levels of that location, e.g. red (sparse), amber (a few) or green (lots). Children must justify their reasoning.

PLENARY

Process the information collected. As a whole group, reflect on the numbers in the findings. Do all groups agree on the RAG rating for each area? What are their conclusions about the animal biodiversity of the setting?

To reflect on the session, pairs of children can consider what they have found out by telling another pair: 3 things they learned from doing this activity, 2 things they want to know more about and 1 question they would like the answer to.

EVALUATION/FOLLOW ON

Encourage the children to explain what they have found out about the diversity of plants in the area. Do the groups agree? Were they surprised by their findings? What were the challenges of this investigation and how were these overcome? What challenges do plants have and how are these overcome? What do they now know about sustainability and the impact of climate change that they didn't know before? Who will they share this new learning with? Is there any action that needs to be taken? What went well and why? What didn't go as well as expected? What could be changed? Who stood out and why?

Back in the classroom

- Interpret, display and compare the data that has been collected. What have they found out and how will they share this information? Does it differ from previous data that may have been collected?

- Using computing, write a non-chronological report of their investigation, or a 'how-to guide' for future classes doing this survey. Plot the findings at the control points on an e-map so that accurate data can be collected over time from the same location.

- Follow up on the plenary activity by finding out more about the 2 'things' and develop a plan to find out the answer to their 1 question.

- Find out how to encourage more biodiversity in a garden through planting, including the difference in ground quality.

PREPARATION

Access the maps of the outside area used in Progressions 1 and 2.

Resources

- Large A3 maps of the area with key features identified from Progressions 1 and 2
- Clipboards, pencils, paper (or sketchbooks) and coloured pencils
- Big question postcards
- Cameras or tablets

Previous learning

Children will have previously used the maps to explore the outdoor area in Progressions 1 and 2.

Children will have a clear understanding of the importance of biodiversity.

LESSON OBJECTIVES

To investigate possible reasons for the range of biodiversity of the outdoor area.

To understand the impact of difference as a factor in biodiversity.

To understand the importance and role of biodiversity in maintaining ecosystems responding to climate change.

National Curriculum content

- Year 5 Science: Study and raise questions about their local environment throughout the year, asking pertinent questions and suggesting reasons for similarities and differences. Describe the differences in the life cycles of a mammal, an amphibian, an insect and a bird. Describe the life process of reproduction in some animals.
- Year 6 Science: Describe how living things are classified into broad groups according to common, observable characteristics based on similarities and differences. Give reasons for classifying animals.
- UKS2 Science: Work scientifically by planning different types of scientific enquiries to answer questions, recording data and reporting and presenting findings.
- KS2 Geography: Use fieldwork to observe, measure and record the human and physical features of the local area using a range of methods, including sketch maps.

CONSIDER

Health and safety

Assess and evaluate hazards and risks within your setting, including any harmful plants, water sources and hazardous boundaries. Consider weather and allergies too.

Equal opportunities or differentiation

Consider how changing or adapting the space, equipment or adult support can benefit or add challenge to the activities.

INTRODUCTION

Explain that in this session, the children will be investigating why some areas of the school grounds support more biodiversity than others. Remind the children that the stronger the ecosystem is, the less of an effect changes (like those relating to climate change) will have.

In Progressions 1 and 2, the children outlined 3 things they learned from doing the survey, 2 things they want to know more about and 1 question they would like the answer to. In pairs or smaller groups, allow time to share this information with new partners. Allow time for the children to revisit areas of the setting to verify information and satisfy curiosity, before re-forming as a class.

WARM-UP ACTIVITY (WHOLE CLASS)

Four corners

This game helps to assess existing scientific knowledge and develop learning. Identify four corners of the space and label them as:

- definitely agree
- definitely disagree
- agree a bit, but...
- disagree a bit, but...

Make a statement. Children move (walk or run) to the appropriate 'corner' to show how much they agree or disagree, giving reasons for their choices to a partner. Play this game by making a series of statements, including correct, incorrect and contentious statements. For example, all invertebrates are insects, all animals like dark places, bees don't need water, etc.

MAIN ACTIVITIES

As a whole group, revise and discuss findings from Progressions 1 and 2 and reason what the children might find in this progression. Use the following questions as conversation starters:

1. What range of environments and habitats were found in the school survey that supported plant, insect and animal life?

2. What research have they done in the classroom to understand what creates the differences in biodiversity? (E.g. habitats, ground quality, shade, moisture, food sources, or use of chemicals.)

3. What can they learn from areas rich in biodiversity to improve other areas that are not thriving?

Challenge 1 (whole class then in pairs)

Big questions

This challenge focuses on the question: why are some areas of the school grounds more biodiverse than others? Share the setting maps from Progressions 1 and 2, discussing the possible reasons for the differing RAG ratings. Ask the children to come up with their own suggestions (use the questions in Appendix 12 as conversation starters).

Then, using big question postcards, ask the children to write their own questions that they would like to find the answers to.

Challenge 2 (in pairs or groups of four)

Revisiting sites

Using the maps of the setting, pairs or small groups revisit each area, specifically focusing on looking for reasons for the differences in biodiversity RAG ratings. At each location, note the time, date and location. Discuss features and characteristics of the locations, looking for possible reasons for the RAG rating (see Appendix 13). Make notes and drawings or take photographs of each setting, while recording answers to the questions in their journals. Add further questions to the big question postcards.

PLENARY

To reflect on the session, remind the children of the main focus question: why do some areas of the setting have greater biodiversity than others? Reinforce the purpose of the session, which was to understand the importance and role of biodiversity in maintaining ecosystems responding to climate change.

Ask the children to share with another pair or group three things they learned from doing this activity. Then ask the children to share any big questions generated during the session. As a class, try to answer each other's questions. Link to the spoken language statutory requirements by encouraging the children to listen and respond appropriately to their peers, giving well-structured descriptions and explanations.

EVALUATION/FOLLOW ON

How well did the children apply their findings? What were the challenges of this investigation and how were these overcome? What went well and why? What didn't go as well as expected? What could be changed? Who stood out and why?

Back in the classroom

- Using computing, write a report about the continuing investigation to propose reasons or hypotheses for possible differences in biodiversity. What do they now know about sustainability and the impact of climate change that they didn't know before? Who will they share this new learning with? Is there any action that needs to be taken urgently?

- Follow up on the plenary activity by finding answers to questions that have arisen.

Progression 4: Revisit
How can we study interactions, connections and interdependence?

PREPARATION

Access the maps of the outside area used in Progressions 1, 2 and 3.

Resources

- Large A3 maps of the area with key features identified from Progressions 1, 2 and 3
- Clipboards, pencils, paper (or sketchbooks) and coloured pencils
- Ball of wool or string
- Large leaves
- A variety of seeds e.g. those naturally occurring in the setting or bought seeds like corn, fennel, radish, lettuce, peas and flower seeds
- Sticky strips or small bags

Previous learning

The children will have previously used maps to explore the outdoor area to identify animals and plants.

In Years 3 and 4, children will have defined biodiversity, explored plants and living things and their habitats, parts of a plant and their function and requirements of plants for life and growth.

LESSON OBJECTIVES

To explore the phenology and life cycles of plants and responses to climate change.

To consider how to increase the range of biodiversity of the outdoor area.

To understand the importance and role of biodiversity in maintaining ecosystems responding to climate change.

National Curriculum content

- Year 5 Science: Study and raise questions about their local environment throughout the year, asking pertinent questions and suggesting reasons for similarities and differences. Describe the differences in the life cycles of a mammal, an amphibian, an insect and a bird. Describe the life process of reproduction in some animals.
- Year 6 Science: Describe how living things are classified into broad groups according to common, observable characteristics based on similarities and differences. Give reasons for classifying animals.
- UKS2 Science: Work scientifically by planning different types of scientific enquiries to answer questions, taking measurements, recording data and reporting and presenting findings.
- KS2 Geography: Use fieldwork to observe, measure and record the human and physical features of the local area using a range of methods, including sketch maps.

CONSIDER

Health and safety

Assess and evaluate hazards and risks within your setting, including any harmful plants, water sources and hazardous boundaries. Consider weather and allergies too.

Equal opportunities or differentiation

Consider how changing or adapting the space, equipment or adult support can benefit or add challenge to the activities.

INTRODUCTION

Explain that the children will be exploring the interactions between plants and animals that promote biodiversity. Allow some time for partner talk to discuss this and come up with some examples. Give an example of how a flower might attract an insect that will then attract its predator, such as a spider, ladybird or wasp. How might climate change be impacting this e.g. plants flowering earlier?

Warm-up activity (whole class)

An interdependent worldwide web

Remind the children that all living things get their energy from food. Green plants use energy from the Sun to make their food. Animals get their energy from eating plants or other animals. This activity will show how we are all interconnected. To reinforce the importance of plants in connections and interdependence in nature, play this web game using a ball of wool (or string).

Gather the group in a circle, show them the wool and explain that you are going to build a school grounds' food web. The wool (representing the flow of energy) should be in the centre of the circle. Each child chooses to be a:

- producer (a plant that gets its energy from the Sun)

- primary consumer (herbivore)

- secondary consumer (carnivore)

- top predator.

Ask anyone who chose the producer (plant) to raise their hand. Starting from the string end, pass the ball around only those children and ask them to hold it tightly with one hand.

Next, ask those who chose primary consumers (herbivores) to raise their hands. Pass the ball of wool to them. Continue with the same process with secondary consumers (carnivores) until you arrive at the top predators and everyone is holding the string. This is your food web and shows how all the species are interdependent.

To further demonstrate this, pose a scenario e.g. 'gardeners came and sprayed herbicides on all the flowers because they thought they were weeds.' All the producers (plants) holding the string now start to gently shake it. Any other living things that feel the shake must then begin to shake the string too. This demonstrates that damage to one part of a food web has an impact throughout.

What happens when we remove a link in the ecosystem? Are the changes more dramatic when the system has many parts or when it has fewer parts? What can we say about the relationships and how stable they are? The more complex or diverse, the more stable it is and the less likely to be impacted by changes.

MAIN ACTIVITIES

Revise previous learning and research about optimum conditions for plants to reproduce and grow. How might changes occur as the impact of climate change manifests, e.g. extended flowering periods, earlier germination or evolutionary change, and what might the impact of this be on food chains? Revise why it is important to have a variety of species in an area. Ask the children:

1. Is it possible to increase biodiversity?

2. What can we do to increase biodiversity across the setting?

Challenge 1 (groups of five)

The life cycle of a flowering plant

In groups of 5, collect a large leaf each. Using marker pens and working as a team, draw a part of the life cycle of a flowering plant on each leaf, arranging them to illustrate the whole life cycle e.g. the seed, germination, growth, reproduction, pollination and seed spreading stages. Explain that this is an example of sexual reproduction (producing seeds). The plant will be similar but not identical to the parent plants. Also, remind the children that some plants reproduce asexually (producing clones of themselves). Taking cuttings of plants is also a way to clone them.

Revise conditions for optimum growth. What do seeds need to germinate? Do all plants need good or moist soil to grow in? Thinking about the setting, what evidence is there of 'germination-friendly' sites in the outdoor learning area? E.g. light, moisture, etc.

Challenge 2 (in pairs or groups of four)

Is there evidence of seeding plants?

Using knowledge gained in previous research, encourage the children to go on a seed hunt. They should collect seeds on sticky strips or in small bags, labelling them with the name of the host plant.

Collect a variety of seeds, label and classify them. These might include wind-dispersed (helicopters, parachutes and grass seeds), bird-dispersed (berries) or animal-dispersed (burs and nuts) seeds. If seeds are not available (time of year dependent), use corn, fennel, radish, lettuce, pea and flower seeds.

Challenge 3 (in pairs or groups of four)

Are there rewilding opportunities?

Raise questions such as: are there any improvements that could be made to the area to attract more insects or promote plant growth? Are there any areas that could be repurposed and seeded, e.g. to create a vegetable patch or mini meadow? What do some areas have that others don't, e.g. green spaces? What are the best locations to grow plants?

The children visit each control point, making notes, sketching out or planning possible improvements or changes to promote biodiversity. This should include increasing the variety of habitats or other planting, based on what they know about conditions for plant growth (note any changes that can and cannot be made, such as creating sheltered areas, making boggy areas, improving drainage, or changing pathways). This might include special places for the children, such as dens, a sensory garden or a quiet area with lots of insects to attract bats and birds.

PLENARY

To reflect on the session, in pairs ask the children to consider what they have found out by telling another pair: 3 things they learned from doing this activity, 2 things they want to know more about and 1 question they would like the answer to.

EVALUATION/FOLLOW ON

What were the challenges of this investigation and how were these overcome? What provided a 'wow' moment? What went well and why? What didn't go as well as expected? What could be changed? Who stood out and why? What do they now know about sustainability and the impact of climate change that they didn't know before? Who will they share this new learning with?

Back in the classroom

- Conduct experiments using some of the seeds to see what they need to grow and thrive. Pose questions to investigate and devise fair tests (see Appendix 14).

- Using computing, write a non-chronological report of the continuing investigation.

- Follow up on the plenary activity by finding answers to questions that have arisen.

- Find out more about soil quality and classification based on its composition.

- Dissect a flower, identifying its parts by placing each part on a sticky strip or drawing from direct observation. Name and label the purpose of each part.

- Extension: Research how to make compost and simple fertiliser.

PREPARATION

Access the maps of the outside area used in Progressions 1-4.

Resources

- Large A3 maps of the area with key features identified from Progressions 1-4
- Soft sponge balls (ideally blue)
- Clipboards, paper and pencils (sketchbooks)
- pH rating soil test kit
- Cameras or tablets
- Magnifying glasses
- Glass jars
- Compost, soil, small pots for planting, sticks for labelling pots
- Seeds, seedlings and runners

Previous learning

In Years 3 and 4, children will have defined biodiversity, explored plants and living things and their habitats, parts of a plant and their function and requirements of plants for life and growth.

Children will have researched differences in soil types based on their texture, the size of the particles and pH levels and how this affects the ground quality.

LESSON OBJECTIVES

To consider the impact of differences in ground quality as a factor in biodiversity.

To follow enquiry-led experimentation with planting.

National Curriculum content

- Year 5 Science: Compare and group together everyday materials on the basis of their properties, including their hardness, solubility and transparency. Know that some materials will dissolve in liquid to form a solution.

- Year 5 Science: Try to grow new plants from different parts of the parent plant. Describe the life process of reproduction in some plants, including sexual and asexual reproduction in plants.

- Year 6 Science: Describe how living things are classified into broad groups according to common, observable characteristics based on similarities and differences.

- Year 6 Science: Recognise that living things have changed over time and that living things produce offspring of the same kind, but normally offspring vary and are not identical to their parents.

CONSIDER

Health and safety

Assess and evaluate hazards and risks within your setting, including any harmful plants, water sources and hazardous boundaries. Consider weather and allergies too.

Equal opportunities or differentiation

Consider how changing or adapting the space, equipment or adult support can benefit or add challenge to the activities.

INTRODUCTION

'Soil is one of our most important natural resources. It is the planet's skin, a rich and complex ecosystem that provides the life systems we all need to survive - oxygen, clean water and food. It is no exaggeration to say that civilisations rise and fall according to the health of the soils on which they are built' (Payton, 2021, p. 4).

Read the above quotation and discuss its meaning. Do the children agree? The report explains how healthy soils are essential to maintaining effective ecosystems, meaning having healthy soils can help mitigate climate change.

Explain that, today, the children will be planning investigations into how the soil in the setting varies and might be a reason for differing plant-growth RAG ratings across the setting.

Revise findings from the previous progressions. What environments and habitats were found in the school survey that promoted insect and animal life or otherwise? What do they think influenced where plants and animals were found? What can we learn from areas rich in biodiversity to improve other areas that are not thriving? Consider the relationship between climate and geology in creating soil e.g. the addition of organic matter such as compost can mitigate the geological state and climatic factors.

Warm-up activity (in groups of 6 or 7)

Soil type game

Revise previous learning about the classification of soil types, explaining the implications of soil type in relation to climate change.

- Sand particles are quite gritty and grainy, and allow water to drain away, but this means water and nutrients are quickly lost.

- Silt has smaller particles than sand and is like flour when dry. It does hold on to water and nutrients, but it also allows some water to drain away.

- Clay has quite dense particles and does not allow water to drain away or travel through it, leading to poor drainage and flooding.

To illustrate how water travels through the different soil types, adapt this playground dodgeball game. Each group has a selection of soft sponge balls (you can choose blue ones to represent water). Choose two group members to be ball throwers. The rest of the group stand side by side to model the different soil types.

- Type one is sand. The group stand loosely apart, with arms reaching out to the side and legs apart. The balls are thrown and pass through the gaps.

- Type two is silt. The group make the gaps slightly smaller, perhaps bending the arms and sidling closer together. The balls are thrown and only some of the balls get through the gaps.

- Type three is clay. The group stands close together so that when the balls are thrown, they cannot pass through.

- Can the children see now how soil type affects drainage?

MAIN ACTIVITIES

Explain that the children will be revisiting the RAG areas of the setting to specifically investigate differences in soil and ground quality, collecting data at each location on the map before undertaking a planting experiment. You may like to allocate the challenges to specific groups to then feedback and present their findings.

Revise conditions for optimum plant growth. What might help plants to grow faster or stronger? What evidence is there, of 'germination-friendly' sites in the outdoor learning area and how does this influence what grows there?

Challenge 1 (pairs)

Does soil pH make a difference in growing conditions?

Using their previous research, ask the children how soil pH might affect what can grow and thrive in specific locations in the setting. Allow some time for partner talk to discuss this and come up with further questions and possible answers.

Demonstrate how to complete a pH rating soil test, following the instructions on the kit, and show the results. Is the soil result acidic, alkaline or neutral? The children then collect the soil sample pH data at identified control points, labelling samples with time, date and location (using the maps to identify the exact location). Can they make links to what planting they find in these locations? Can they generate a hypothesis about the difference pH makes to planting?

Challenge 2 (in pairs or groups of four)

How does soil type vary in different locations?

Remind the children of the warm-up game about soil types – sand, silt and clay. The children are now going to perform sediment tests to demonstrate the composition of soil at specific locations (i.e. the relative amounts of sand, silt and clay in a soil sample).

How to conduct a soil sediment test:

Take a soil sample from specific locations in the setting, marking the map to identify the location. This can be supported by taking a photograph or drawing and labelling the soil's initial appearance before placing the sample in a glass jar, labelled with time, date and location.

Pour water into the jar (not to the top) and put on the lid.

Shake the jar to saturate the soil sample and put it to one side to settle (24 hours).

Notice and record changes in the appearance of the soil over time.

Observe the differences in the layers of the samples. What does this indicate about the composition of the soil? What proportions are there, e.g. 30% clay, 40% sand and 30% silt? How will they present their findings?

Challenge 3 (in groups)

Does soil quality and type make a difference in plant growth?

Using the research about which plants thrive in different soils, the children should devise and carry out experiments on planting in different soil types and conditions. They should use seeds that grow into plants that are similar but not identical to the parent plant (sexual reproduction) or take cuttings and plant runners to grow duplicates or clones identical to the parent plant (asexual reproduction).

Planting in small pots, the children should devise and carry out experiments to test a hypothesis, for example, planting in different soil types.

The children should carefully label the pots, ensuring that the data is recorded and accurate, e.g. date, time, location, weather conditions (if planted or potted outdoors), amount of soil, type of soil and amount of water provided. The plants should then be monitored. Take photographs as evidence of each stage of growth. Finally, plan how to present the findings.

PLENARY

To reflect on the session, in pairs ask the children to consider what they have found out by telling another pair: 3 things they learned from doing this activity, 2 things they want to know more about and 1 question they would like the answer to.

EVALUATION/FOLLOW ON

What were the challenges of these investigations and how were these overcome? What went well and why? What didn't go as well as expected? What could be changed? Who stood out and why? What do they now know about sustainability and the impact of climate change that they didn't know before? Who will they share this new learning with? Is there any action that needs to be taken?

Additional enquiry-led challenges might include: Does soil temperature affect plant growth? Does being near a wall affect plant growth? What difference does the 'time of year' make to plant growth and optimum growing conditions?

Back in the classroom

- In preparation for the next progression, research planting for rewilding the school grounds to improve biodiversity.

- Research how to make compost and simple fertiliser.

- Follow up on the plenary activity by finding answers to questions that have arisen.

PREPARATION

- Assess access to grounds and get agreements/permissions.
- This progression builds on the RE units in Bloomsbury's *The National Curriculum Outdoors: Year 6* – consider completing this first.
- Access the maps of the outside area used in Progressions 1-5.

Resources

- Large A3 maps of the area with key features identified from Progressions 1-5
- Gardening equipment
- Plants, seeds, stem and root cuttings, tubers, bulbs, seedlings or runners
- Clipboards, pencils, paper (or sketchbooks) and coloured pencils
- Species ID App, identification charts or classification keys
- Cameras or tablets

Previous learning

- Children will have researched soil and ground quality and have an understanding of environments to promote growth and biodiversity.
- Children will have found out how national parks plan and promote sustainability.

LESSON OBJECTIVES

To use what we know to plan to increase the range of biodiversity of the outdoor area and develop a micro national park on the school grounds.

To understand the importance and role of biodiversity in maintaining ecosystems in response to climate change.

National Curriculum content

- Year 5 Science: Study and raise questions about their local environment throughout the year, asking pertinent questions and suggesting reasons for similarities and differences. Describe the differences in the life cycles of a mammal, an amphibian, an insect and a bird. Describe the life process of reproduction in some animals.

- Year 6 Science: Describe how living things are classified into broad groups according to common, observable characteristics based on similarities and differences. Give reasons for classifying animals.

- UKS2 Science: Work scientifically by planning different types of scientific enquiries to answer questions, recording data and reporting and presenting findings.

CONSIDER

Health and safety

Assess and evaluate hazards and risks within your setting, including any harmful plants, water sources and hazardous boundaries. Consider weather and allergies too.

Equal opportunities or differentiation

Consider how changing or adapting the space, equipment or adult support can benefit or add challenge to the activities.

INTRODUCTION

Explain that the children will be using what they have learned to plan and plant to improve the biodiversity of the setting. Why is this important in relation to climate change? What facts do they now know about sustainability and the impact of climate change that they will be applying today in their transformation of the school grounds?

The improvements that they make will transform the setting into a micro national park that they will manage. Share what they have found out about national parks, management, stewardship, management structures, etc. Allow some time for partner talk to discuss this and come up with some examples. How can this 'national park' structure be applied to the school setting, e.g. improving biodiversity and taking action on sustainability in the face of climate change?

Warm-up activity (individual)

Nature connection

Lead a mindfulness activity to raise awareness of the surroundings and promote a nature connection. Ask the children to find an individual sit spot where they feel comfortable and take a minute to tune in to what is around them. Sit quietly and connect to nature.

Ask them: What do you hear? What is the closest sound? Is it a natural or manmade sound? Can you identify sounds further away? Are there pleasing or displeasing sounds? Identify where each sound is being created. Reflect on the powerful feelings that being in nature creates.

As the children return to the whole group, ask them to quietly think about how they are taking action in their setting. They are caring for nature and doing something to protect it.

MAIN ACTIVITIES

In pairs, ask the children to discuss, evaluate and share the planting progress from the previous progressions, including information about their experiment and the impact of their control measures. Ask the following questions to prompt their discussions:

How have the seeds, seedlings and cuttings responded to the control measures?

What are the main differences between thriving areas and areas that are sparsely populated by plants or animals?

How can conditions for growth be replicated?

Ask the children how can they learn from the experiments and improve the biodiversity of the outdoor area to mitigate the effects of climate change. The children can come up with their own hypotheses and questions based on what they know from their research, revising conditions for optimum growth, and sharing what animals, birds and insects need to thrive in the face of climate change. In discussion, the children are empowered to make a difference in relation to climate change and sustainability.

Challenge 1 (whole class then in groups)

Plan and take action for planting

Ask the children to identify an area of ground that has been graded red or amber for plant biodiversity that could be used to cultivate new plants. These might include raised beds, unused areas, re-purposed areas or using flowerpots on hard-standing or paved areas.

Decide as a class on the growing conditions for seeds, stem and root cuttings, tubers, bulbs, seedlings and runners. Plan the planting process and how changes in climate might impact these conditions. What preventative action can be taken? Working in groups, prepare the ground or pots, providing food such as fertilizer (which they may make themselves), creating compost heaps and bins and finding ways to capture and store water. Ensure that the processes are recorded scientifically. For quick results, plant fast germinating seeds, such as a wallflower, sunflower, peas, beans or radishes, pot strawberry runners or take cuttings from plants such as geraniums.

Challenge 2 (in pairs or groups of four)

Take action for sustainability

Consider sustainability in relation to the project to mitigate climate change effects. What installations might be placed in the setting to encourage insects, birds and animals?

How might growing conditions be enhanced:

- by using cold frames or warm beds

- in a greenhouse or polytunnel

- using solar panels to create power for warmth or to run a water pump or filter?

Visit each RAG area, itemising what might be practicable at each site, and sketching the installation to support persuasive writing back in the classroom.

Challenge 3 (in groups)

Take action for rewilding

At each control point, or at points otherwise identified during the unit, make a plan to develop and improve rewilding (based on what they now know about conditions for plants to thrive), where necessary. Initially explored in Progression 4, Challenge 3.

PLENARY

Remind the children that this is an ongoing project to be managed by the children themselves as part of a pupil action group (supported by the whole school, including staff). The children will evaluate and share their plans and what they have enacted. What future actions need to be taken to ensure the sustainability of the project? Decide when would be best to evaluate the success of the biodiversity project, which is intended to be ongoing, sustainable and evaluated regularly throughout the school year(s).

As part of a wider project to develop the grounds as a micro national park, take the time to visit and sit in the areas, reflecting on a long-term plan with termly targets. This should include evaluation and next steps planning.

EVALUATION/FOLLOW ON

What were the challenges of this investigation and how were these overcome? What do the children now know about sustainability and the impact of climate change that they didn't know before? Who will they share this new learning with? Is there any ongoing action that needs to be taken? How can what they have learned be applied to other areas? What went well and why? What didn't go as well as expected? What could be changed? What might you do differently next time? Who stood out and why?

Have children raised questions about the diversity of their local environment? Have they recorded data and critically reflected and reported their results? Have they used the results of their investigations to develop specific areas of the school grounds to encourage growth and biodiversity in the outdoor learning area as part of the longer-term project of hope and positivity?

Back in the classroom

- Form a management council to maintain the grounds. Celebrate and publicise what has been done in an outdoor assembly focusing on sustainability and stewardship, inviting parents and carers. Share information regularly with the immediate and wider community.

- As part of a DT project, make cold frames, bird feeders, nesting boxes, hedgehog homes, bug hotels, rain gauges, weather stations or wind turbines.

- Plan a campaign to fund a polytunnel or greenhouse. Use solar power for decoration, heat or light.

- Create an eco-school badge for the whole school and campaign for regular outdoor lessons throughout the school. Hold a community conference led by accredited climate change champions from each class, perhaps linking this to an eco-school accreditation programme.

Chapter 6
Concluding thoughts and next steps

A brief review of what has been covered in this book

This chapter provides a brief review of the book, and an invitation to take your work further. Key questions facing teachers working in the complex and demanding school workplace are 'Should we...', 'Can we...' and 'How might we...' integrate sustainability, climate change and biodiversity (SCCB), particularly in outdoor contexts, in our schools? We hope that this book helps to provide answers in each case. The new Sustainability and Climate Change Strategy (Department for Education, 2022) should go a long way towards legitimising this agenda for change (or rather these agendas plural, given the inextricable nature of these issues). These will, of course, need to be set alongside other competing agendas and priorities. The case might yet need to be strengthened to fully convince school policy-makers and decision-makers to adopt and embed an outdoor approach to SCCB education, so further justifications might be needed to bolster the argument.

Chapter 1 sought to provide such further justification pertinent to the goals of schools and education. The case was made on intellectual, societal/citizenship and positive psychological grounds, enhancing learners' critical thinking and problem solving, especially if driven by an enquiry-led approach to learning. Similarly, working collaboratively and in a pro-environmental fashion will enhance competencies as environmental citizens and stewards of the environment. This will also enhance self-esteem and self-efficacy and provide a sense of meaning and belonging (crucial dimensions of positive psychology). Another dimension of positive psychology highlighted is the proven positive health benefits – both physical and mental – of being outdoors and in nature. Similarly, experiential engagement with familiar local environments – school grounds and nearby nature – is more likely to engender positive place attachment and nature connection (crucial further dimensions of healthy flourishing).

Making the case on educational and wider factors is just the beginning. Next comes the challenging practical question of 'How can we...' incorporate SCCB through outdoor learning into our existing contexts and practice? Teachers are naturally creative and are able to translate policy into practice, however, it can sometimes be difficult to get started. This is where the main body of this book comes in.

The book provides inspirational and practical guidance with exemplars to empower teachers and schools to creatively shape change in their own contexts. The idea has been not to be overly prescriptive since no two school contexts are the same and the rapid changes in our world require educators to adopt a flexible approach to teaching and learning. It is quite likely that the policy context, including the National Curriculum, will change in the next few years. The ideas presented in this book will retain relevance as they are intended to illustrate how aspects of the curriculum can be delivered outdoors while contributing to SCCB education and providing opportunities for personal and social development and citizenship.

Chapter 2 provides the all-important focus on leadership to drive the change at a variety of scales, from curricular subjects to whole-school culture, policy and plans. This chapter provided a cyclical model (Figure 2.1), which can provide a powerful framework to structure innovative transformational leadership. The emphasis is on achievable change over time, and importantly, promoting a form of distributed ownership and leadership to include both staff and pupils. The model starts with critical reflection on developing a shared and collective ethos, mindset and vision and progresses through to practical considerations of impact. Along the way, there is the crucial consideration of the necessary 'upskilling' of the school community staff and pupils to empower them as positive agents of change. The recommendation is that such efforts lead to a long-term sustainability plan for the school. The chapter brings the model to life with inspiring real-world anecdotes, reflections and case studies from practitioners.

Chapter 3 focuses specifically on the school's spatial and geographical context in terms of the school grounds and the home locality it serves – the 'nearby'. Strategically, these represent the most promising locations for appropriate outdoor learning by evading the typically cited barriers of cost (both financially and in time to plan and execute), risk (and associated health and safety considerations) and difficulty (with many teachers lacking confidence). The local area provides the most important context to effect meaningful, demonstrable and often immediate change that drives engagement and motivation. Thus, small-scale habitat creation or regeneration in the school grounds across the whole site can and will make a real impact on mitigating climate change and regenerating biodiversity at international and even global levels. Thus, learners 'acting locally, but thinking globally' will be able to recognise and exercise their agency as 'global citizens'. The local area represents a context that facilitates school-community links and engenders meaningful and long-lasting bonds of affection with the locality, and the people and nature it comprises. Chapter 3 provides helpful advice on how to approach the 'greening' of the school grounds, and how to recognise and overcome potential barriers and issues. In particular, it provides a user-friendly set of principles and recommendations. These can be used to inform the various phases of the leadership model presented in Chapter 2. Once again, real-world examples are provided to demonstrate potential.

Chapters 4 and 5 present exemplar units of work for lower and upper Key Stage 2 respectively. Taken together, these chapters demonstrate progression in learning, knowledge and skills across the Key Stages. Whilst the materials foreground the delivery of the National Curriculum Science programmes of study, they are very much framed in an interdisciplinary manner, with a strong emphasis on developing competencies demanded of global citizens in the 21st century. Thus, the units emphasise enquiry-based learning and opportunities to develop critical thinking and communication skills, as well as technology literacy. There are opportunities for the children to work independently and in small groups or pairs (with adults providing positive support), in addition to whole-class teacher-directed learning. The units advocate regular and sustained use of the outdoors and across the seasons to experience phenological changes and deepen nature connection. The units are presented in a form that could be applied with minimal adaptation in most, if not all, contexts. Equally, they could represent the starting point for teachers to creatively adapt and take ownership in relation to their own specific contexts. Perhaps most importantly, both chapters present teaching and learning ideas that are fun and engaging for pupils and teachers alike.

An invitation – what will be your next steps?

The introduction presented the idea of the unique knowledge blend that is associated with educators, combining knowledge of a topic (Content Knowledge or CK) with knowledge of how to teach (Pedagogical Knowledge, PK) to arrive at the best strategies to teach a topic to specific learners within a particular context (Pedagogical Content Knowledge or PCK). Whilst we acknowledge that teachers are already experts in PCK in the context of classroom-based teaching, we hope that this book will provide stimulus to encourage teaching and learning about SCCB in an outdoor context (i.e. Outdoor Pedagogical Knowledge or OPK).

The first chapter sought to introduce the overarching concept of sustainability and the interrelated themes of climate change and biodiversity thereby enhancing teachers' CK (what might be termed SCCB Content Knowledge or SCCBCK). However, this has only scratched the surface. Sustainability represents a very broad and complex concept that incorporates a number of highly interconnected themes and issues, many quite contentious. There are currently 17 interrelated Sustainable Development Goals (SDGs) recognised by the United Nations (see Figure 1.1). In this book, we implicitly advocate for all of the goals, including 3: Good Health and Wellbeing; 4: Quality Education; 5: Gender Equality; and 10: Reduced Inequality.

However, we could not possibly have attempted to do justice to all of these SDGs nor indeed all aspects that fall under the umbrella of sustainability. Fortunately, there is an increasing range of resources that are available to provide teachers with support in terms of the broader sustainability agenda (for examples see Scoffham and Rawlinson, 2022 or UNESCO, 2020). Equally, climate change and biodiversity are both themes that are potentially vast in their own right and warrant much further exploration than has been possible in Chapter 1. Consequently, a 'next step' is for teachers to deepen their SCCBCK and explore their inextricable interconnections.

The introduction also posed the notion of the necessary PK and PCK that relate to teaching outdoors (OPK). Throughout we have implicitly emphasised certain approaches such as 'enquiry-based learning' and 'place-based learning'. We would encourage practitioners to explore these innovative approaches to teaching and learning, as teaching outdoors requires different pedagogical knowledge.

Thus, you might be inspired to undertake further professional development through CPD or at postgraduate level to develop your SCCBCK and/or OPK. You might also consider joining existing Communities of Practice or Research Interest Groups or, indeed, set one up yourself. This could involve like-minded colleagues and even students within your school setting (or across schools). More ambitiously, it could aim to also connect with outdoor learning practitioners and volunteers in the locality and beyond. But perhaps the most rewarding 'next step' would be to attempt to put the ideas presented in this book into practice in your own setting – to develop and enact your own PCK in relation to teaching SCCB outdoors. As the model introduced in Chapter 2 emphasises, this is an ongoing process that is likely to involve a number of stages and a collaborative and distributed effort from staff and, ideally, pupils.

Key questions for practice:

- How might you adopt, adapt or recreate the units and progressions presented in this book to make them more place-responsive to the particular affordances, opportunities, challenges and issues presented in your school, home locality and outdoors nearby?

- What are the best ways for adults and children to work together? Can the traditional roles be reversed, with learners taking the lead in decision-making and mentoring? What are the opportunities for genuine 'intergenerational learning'?

Typically, the role of a mentor is associated with adults and might be seen to include:

- modelling the activity

- acknowledging and overcoming challenges

- encouraging questioning and wider speculation

- posing open questions to facilitate making connections and relationships

- modelling and assisting with accurate identification and recording

- providing technical assistance where required

- supervising safety.

How confident do you feel in each or all of these?

With experience, there is no real reason why these mentorship roles cannot increasingly be adopted by the learners themselves. This can be both to lead their peers and adults, including teachers and members of the wider community. How might you shift your practice to nurture your learners as 'environmental mentors' in your school?

We wish you well on what will no doubt be a challenging but very rewarding professional journey.

Author biographies

Dr. Alun Morgan is Senior Lecturer in Education at the University of Plymouth where he leads courses on Environmental and Sustainability Education and Outdoor Learning. He has over 30 years of experience in education including as a school teacher, advisor, lecturer and researcher.

Deborah Lambert has a wealth of experience as a school teacher and lecturer in further education. Deborah works with schools across the country, offering curriculum-linked outdoor learning sessions with classes and training and support for teachers. She led the international Erasmus+ Natural Schooling project and is chair of the Natural Schooling Research Interest Group.

Michelle Roberts is an Assistant Headteacher, Sports Partnership Director and a consultant in PE and outdoor learning. Michelle has a wealth of experience working in primary schools across the south west of England providing continuous professional development in outdoor leadership for teachers, and site development. The successful Wild Tribe outdoor learning programme has underpinned this work, and the newly developed Earth Tribe supports the development of a sustainable curriculum.

Sue Waite, a former primary schoolteacher and Associate Professor in Outdoor Learning at the University of Plymouth, is a member of the International School Grounds Alliance Leadership group. She led Natural Connections, an ambitious UK demonstration project about curriculum learning in natural environments, funded by Natural England, DEFRA and Historic England.

Further reading

Additional information about testing and resulting planting can be found here:

BBC Gardeners' World Magazine (2019). How to test your soil pH. [online] Available at: https://www.gardenersworld.com/how-to/maintain-the-garden/how-to-test-your-soil-ph/ [Accessed 22 Sep. 2023].

Droxford Junior School's Landscape Strategy from Chapter 2 can be found here:

Droxford Junior School. (2023). Outdoor Learning and Play: Landscape Strategy. [online] Available at: https://www.droxfordjunior.co.uk/page/?title=Outdoor+Learning+and+Play&pid=26 [Accessed 22 Sep. 2023].

More information on the five pathways to nature connection can be found here:

Lumber, R., Richardson, M. and Sheffield, D. (2017). Beyond knowing nature: Contact, emotion, compassion, meaning, and beauty are pathways to nature connection. PLOS ONE, 12(5). doi:https://doi.org/10.1371/journal.pone.0177186.

Additional case studies and information can be found here:

National Curriculum Outdoors. (n.d.). National Curriculum Outdoors. [online] Available at: https://nationalcurriculumoutdoors.com/climate-change-in-ks2 [Accessed 22 Sep. 2023].

Further details of St Albans School projects in Chapter 2 can be found here:

Newman, J. (2022). Supporting pupil's learning and development through engagement with nature. [online] teaching. blog.gov.uk. Available at: https://teaching.blog.gov.uk/2022/11/11/supporting-pupils-learning-and-development-through-engagement-with-nature/ [Accessed 22 Sep. 2023].

Additional information about testing and resulting planting can be found here:

RHS. (n.d.). Soil: understanding pH and testing soil. [online] Available at: https://www.rhs.org.uk/soil-composts-mulches/ph-and-testing-soil [Accessed 22 Sep. 2023].

Information on a citizen science project to inspire your own can be found here:

Wytham Woods. (n.d.). Citizen Science. [online] Available at: https://www.wythamwoods.ox.ac.uk/citizen-science [Accessed 22 Sep. 2023].

The following articles were published by the Swedish ecosystem services and support the 'strengthening ecosystem services in Swedish preschool and school grounds' case study in Chapter 3:

Almers, E., Askerlund, P. and Samuelsson, T. (2023). The Perfect Schoolyard for Future Children: Primary School Children's Participation in Envisioning Workshops. *Children, Youth and Environments*, 33(1), pp.101–121. doi:https://doi.org/10.1353/cye.2023.0002.

Almers, E., Askerlund, P., Samuelsson, T. and Waite, S. (2021). Children's preferences for schoolyard features and understanding of ecosystem service innovations – a study in five Swedish preschools. *Journal of Adventure Education and Outdoor Learning*, 21(3), pp.230–246. doi:https://doi.org/10.1080/14729679.2020.1773879.

Askerlund, P., Almers, E., Tuvendal, M. and Waite, S. (2022). Growing nature connection through greening schoolyards: preschool teachers' response to ecosystem services innovations. *Education 3-13*. doi:https://doi.org/10.1080/03004279.2022.2148485.

Kjellström, S., Andersson, A.-C. and Samuelsson, T. (2020). Professionals' experiences of using an improvement programme: applying quality improvement work in preschool contexts. *BMJ Open Quality*, 9(3). doi:https://doi.org/10.1136/bmjoq-2020-000933.

Lecusay, R., Mrak, L. and Nilsson, M. (2022). What is Community in Early Childhood Education and Care for Sustainability? Exploring Communities of Learners in Swedish Preschool Provision. *International Journal of Early Childhood*, 54(3), pp.51–74. doi:https://doi.org/10.1007/s13158-021-00311-w.

The following resources are recommended to support further research into green infrastructure for educational settings discussed in Chapter 3:

Children and Nature Network (n.d.). School Ground Greening. [online] Children & Nature Network. Available at: https://www.childrenandnature.org/schoolgroundgreening/ [Accessed 15 Sep. 2023].

Department for Children, Schools and Families (2008). Planning a sustainable school: Driving school improvement through sustainable development acknowledgments. [online] Available at: https://support.rm.com/_rmvirtual/Media/

Downloads/DCSF_Planning_Guide.pdf [Accessed 15 Sep. 2023].

European Union (n.d.). New European Bauhaus Prizes 2023. [online] Available at: https://prizes.new-european-bauhaus.eu/ [Accessed 15 Sep. 2023].

Guardian News (2019). Greta Thunberg and George Monbiot make short film on the climate crisis. YouTube. Available at: https://www.youtube.com/watch?v=-Q0xUXo2zEY [Accessed 15 Sep. 2023].

International School Grounds Alliance (n.d.). About ISGA. [online] ISGA. Available at: https://www.internationalschoolgrounds.org/about [Accessed 15 Sep. 2023].

Natural History Museum (n.d.). Climate change. [online] Nhm.ac.uk. Available at: https://www.nhm.ac.uk/discover/climate-change.html [Accessed 15 Sep. 2023].

Natural History Museum (n.d.). National Education Nature Park and Climate Action Awards. [online] www.nhm.ac.uk. Available at: https://www.nhm.ac.uk/about-us/national-impact/national-education-nature-park-and-climate-action-awards-scheme.html?utm_content=hero-cta&utm_campaign=nature-park&utm_medium=email&utm_source=2330752_ma-enews-generic-naturepark-20230518&dm_i=2XEG [Accessed 15 Sep. 2023].

Nature Friendly Schools (n.d.-c). Our Impact. [online] www.naturefriendlyschools.co.uk. Available at: https://www.naturefriendlyschools.co.uk/our-impact [Accessed 15 Sep. 2023].

RHS (n.d.). RHS Campaign for School Gardening. [online] schoolgardening.rhs.org.uk. Available at: https://schoolgardening.rhs.org.uk/ [Accessed 15 Sep. 2023].

References

Adey, P. (2004). *The professional development of teachers: practice and theory*. Dordrecht: Kluwer Academic Publishers.

Altman, I., Chawla, L. and Low, S.M. (1992). *Place attachment*. New York: Plenum Press, pp.63–86.

Antoniadis, D. N. Katsoulas, N. & Papanastasiou, D.K. (2020) Thermal environment or urban schoolyards: current and future design with respect to children's thermal comfort. *Atmosphere*, 11, 1144, Available at: https://www.mdpi.com/2073-4433/11/11/1144

Austin, S. (2021) 'The school garden in the primary school: Meeting the challenges and reaping the benefits', *Education 3-13*, 50(6), pp. 707–721. doi:10.1080/03004279.2021.1905017.

Balmford, A., Clegg, L., Coulson, T. and Taylor, J. (2002). Why Conservationists Should Heed Pokemon. *Science*, 295(5564), pp.2367–2367.

Bosevska, J. and Kriewaldt, J. (2019) 'Fostering a whole-school approach to sustainability: Learning from one school's journey towards Sustainable Education', *International Research in Geographical and Environmental Education*, 29(1), pp. 55–73. doi:10.1080/10382046.2019.1661127.

Bristol Climate Hub. (n.d.). *Become a citizen scientist*. [online] Available at: https://www.bristolclimatehub.org/climate_actions/become-a-citizen-scientist/ [Accessed 19 Sep. 2023].

Broom, C. (2017). Exploring the Relations Between Childhood Experiences in Nature and Young Adults' Environmental Attitudes and Behaviours. *Australian Journal of Environmental Education*, 33(1), pp.34–47.

Butterfly Conservation. (n.d.). *Big Butterfly Count*. [online] Available at: https://bigbutterflycount.butterfly-conservation.org/ [Accessed 19 Sep. 2023].

Cantell, H., Tolppanen, S., Aarnio-Linnanvuori, E. and Lehtonen, A. (2019). Bicycle model on climate change education: presenting and evaluating a model. *Environmental Education Research*, 25(5), pp.717–731.

CCEA (2007). *The Northern Ireland Curriculum Primary*. [online] ccea.org.uk. Available at: https://ccea.org.uk/downloads/docs/ccea-asset/Curriculum/The%20Northern%20Ireland%20Curriculum%20-%20Primary.pdf [Accessed 13 Sep. 2023].

Children & Nature Network. (n.d.). *School Ground Greening | Children and Nature Network*. [online] Available at: https://www.childrenandnature.org/schoolgroundgreening/#:~:text=Benefits%20of- [Accessed 7 Sep. 2023].

Coads Green Primary School. (n.d.). *School Ethos: Climate Change and Sustainability*. [online] Available at: https://www.coads-green.cornwall.sch.uk/web/climate_change__sustainability/636535 [Accessed 8 Sep. 2023].

Collin J., & Smith E. (2021). *Education Endowment Foundation: Effective professional development guidance report*. [online] Available at: https://educationendowmentfoundation.org.uk/education-evidence/guidance-reports/effective-professional-development

Cornwall Green Schools. (n.d.). *Cornwall Green Schools*. [online] Available at: https://cornwallgreenschools.co.uk/ [Accessed 8 Sep. 2023].

Cremin, T. (2017). *Teaching Outdoors Creatively*. Oxon: Routledge. Series Editor Foreword.

Curriculum for Wales (2022). *HwB* [online] Available at: https://hwb.gov.wales/curriculum-for-wales [Accessed 16 Oct. 2023]

Cutting, R. and Kelly, O. (2014). *Creative teaching in primary science*. London: Sage.

Dasgupta, P. (2021). *The Economics of Biodiversity: The Dasgupta Review*. [online] *GOV.UK*. HM Treasury. Available at: https://assets.publishing.service.gov.uk/government/uploads/system/uploads/attachment_data/file/962785/The_Economics_of_Biodiversity_The_Dasgupta_Review_Full_Report.pdf [Accessed 8 Sep. 2023].

Department for Children, Schools and Families (2008). *Planning a Sustainable School: Driving School Improvement through Sustainable Development*. [online] Available at: https://support.rm.com/_rmvirtual/Media/Downloads/DCSF_Planning_Guide.pdf [Accessed 13 Sep. 2023].

Department for Education (2013a). *National Curriculum*. [online] GOV.UK. Available at: https://www.gov.uk/government/collections/national-curriculum [Accessed 7 Sep. 2023].

Department for Education (2013b). *National curriculum in England: primary curriculum*. [online] GOV.UK. Available at: https://www.gov.uk/government/publications/national-curriculum-in-england-primary-curriculum [Accessed 11 Sep. 2023].

Department for Education (2014). *Area guidelines for mainstream schools*. [online] GOV.UK. Available at: https://assets.publishing.service.gov.uk/government/uploads/system/uploads/attachment_data/file/905692/BB103_Area_Guidelines_for_Mainstream_Schools.pdf [Accessed 13 Sep. 2023].

Department for Education (2014). *Building Bulletin 103: Area Guidelines for mainstream schools*. [online] Available at: https://assets.publishing.service.gov.uk/government/uploads/system/uploads/attachment_data/file/905692/BB103_Area_Guidelines_for_Mainstream_Schools.pdf

Department for Education (2022). *Sustainability and climate change: a strategy for the education and children's services systems*. [online] GOV.UK. Available at: https://www.gov.uk/government/publications/sustainability-and-climate-change-strategy/sustainability-and-climate-change-a-strategy-for-the-education-and-childrens-services-systems [Accessed 5 Sep. 2023].

Department for Education (2023). *Sustainability Leadership and Climate Action Plans in Education*. [online] GOV.UK. Available at: https://www.gov.uk/guidance/sustainability-leadership-and-climate-action-plans-in-education [Accessed: 11 Dec. 2023].

Department for Education and Employment, Qualifications and Curriculum Authority. (1999). *The National Curriculum for England: Science Key Stages 1-4*. London: HMSO.

Department for Education and Skills (2004). *Building Bulletin 99: Briefing Framework for Primary School Projects*. [online] Available at: http://www.educationengland.org.uk/documents/pdfs/2004-building-bulletin-99-pri.pdf

Department for Education and Skills (2006a). *Learning Outside the Classroom Manifesto*. [online] Nottingham: DfES Publications. Available at: https://thegrowingschoolsgarden.org.uk/downloads/lotc-manifesto.pdf [Accessed 7 Sep. 2023].

Department for Education and Skills (2006b). *Sustainable Schools*. [online] nationalarchives.gov.uk. Available at: https://webarchive.nationalarchives.gov.uk/ukgwa/20060718042253/http:/www.dfes.gov.uk/consultations/downloadableDocs/Executive%20Summary%20Final.pdf [Accessed 7 Sep. 2023].

Department for Education and Skills (2007). *Designing School Grounds*. [online] GOV.UK. Available at: https://assets.publishing.service.gov.uk/government/uploads/system/uploads/attachment_data/file/276691/schools_for_the_future_-_designing_school_grounds.pdf [Accessed 7 Sep. 2023].

Department for Environment Food and Rural Affairs (2019). *Gove kicks off year of Green Action*. [online] GOV.UK. Available at: https://www.gov.uk/government/news/gove-kicks-off-year-of-green-action [Accessed 22 Nov. 2023].

Department for Environment Food and Rural Affairs (2019). *Landscapes Review*. [online] GOV.UK. Available at: https://assets.publishing.service.gov.uk/government/uploads/system/uploads/attachment_data/file/833726/landscapes-review-final-report.pdf [Accessed 7 Sep. 2023].

Dewey, J. (1899). *The School and Society*. Chicago: The University Chicago Press.

Droxford Junior School (n.d.). *Droxford Junior School | Home*. [online] Available at: https://www.droxfordjunior.co.uk/ [Accessed 14 Sep. 2023].

Droxford Junior School (n.d.). *School Council and pupil voice*. [online] Available at: https://www.droxfordjunior.co.uk/page/?title=School%2BCouncil%2Band%2BPupil%2BVoice&pid=13 [Accessed 16 Oct. 2023].

Dunlop, L. and Rushton, E.A.C. (2022). Putting climate change at the heart of education: Is England's strategy a placebo for policy? *British Educational Research Journal*, 48(6), pp. 1083–1101.

Education Scotland (2020). *Building your Curriculum: Outside and In*. [online] moray.gov.uk. Available at: http://www.moray.gov.uk/downloads/file89167.pdf [Accessed 14 Sep. 2023].

Evergreen (2011). *Planning & Designing Green School Grounds in Pembina Trails School Division*. [online] Available at: https://sbptsdstor.blob.core.windows.net/media/Default/medialib/planning-and-designing-green-schools.7373b59608.pdf [Accessed 14 Sep. 2023].

Food A Fact Of Life (n.d.-a). *Food origins (7-11 Years) - Food A Fact Of Life*. [online] www.foodafactoflife.org.uk. Available at: https://www.foodafactoflife.org.uk/7-11-years/where-food-comes-from-7-11-years/food-origins-7-11-years/ [Accessed 14 Sep. 2023].

Food A Fact Of Life (n.d.-b). *Growing clubs - Food A Fact Of Life*. [online] www.foodafactoflife.org.uk. Available at: https://www.foodafactoflife.org.uk/whole-school/whole-school-approach/growing-clubs/ [Accessed 14 Sep. 2023].

Glackin, M. and Greer, K. (2021). 'What counts' as climate change education? Perspectives from policy influencers. *School Science Review*, 103(383), pp.15–22.

Green, J. (2015). *The Environmental Curriculum: Opportunities for Environmental Education across the National Curriculum for England Early Years Foundation Stage & Primary*. [online] naee.org.uk. Walsall: NAEE. Available at: https://naee.org.uk/wp-content/uploads/2015/06/NAEE_The_Environmental_Curriculum.pdf [Accessed 7 Sep. 2023].

Green, M. and Rayner, M. (2020). 'School ground pedagogies for enriching children's outdoor learning', *Education 3-13*, 50(2), pp. 238–251. doi:10.1080/03004279.2020.1846578.

Gutiérrez, L.M. (1994). Beyond Coping: An Empowerment Perspective on Stressful Life Events. *Journal of Sociology & Social Welfare*, 21(3), p.201.

Hancock, L. (n.d.). *What is biodiversity and why is it under threat?* [online] World Wildlife Fund. Available at: https://www.worldwildlife.org/pages/what-is-biodiversity [Accessed 19 Sep. 2023].

Harvey, D. (2022). *5 key benefits of outdoor learning*. [online] Learning through Landscapes. Available at:

https://ltl.org.uk/news/5-key-benefits-of-outdoor-learning/ [Accessed 7 Sep. 2023].

Health and Safety Executive (n.d.). *Sensible health and safety management in schools*. [online] www.hse.gov.uk. Available at: https://www.hse.gov.uk/education/sensible-leadership/index.htm [Accessed 15 Sep. 2023].

Hickman, C., Marks, E., Pihkala, P., Clayton, S., Lewandowski, R.E., Mayall, E.E., Wray, B., Mellor, C. and van Susteren, L. (2021). Climate anxiety in children and young people and their beliefs about government responses to climate change: a global survey. *The Lancet Planetary Health*, 5(12).

Hicks, D. (2014). *Educating for hope in troubled times: climate change and the transition to a post-carbon future*. London: Institute Of Education Press.

HM Government (2018). *A Green Future: Our 25 Year Plan to Improve the Environment*. [online] GOV.UK. Available at: https://assets.publishing.service.gov.uk/government/uploads/system/uploads/attachment_data/file/693158/25-year-environment-plan.pdf [Accessed 7 Sep. 2023].

HM Treasury (2021). *Final report – the economics of biodiversity: The Dasgupta Review*. [online] GOV.UK. Available at: https://www.gov.uk/government/publications/final-report-the-economics-of-biodiversity-the-dasgupta-review [Accessed: 22 Nov. 2023].

IPCC (2014). *Ar5 climate change 2014: Mitigation of climate change, IPCC*. [online] Available at: https://www.ipcc.ch/report/ar5/wg3/ [Accessed: 17 October 2023].

IPCC (2021) Summary for Policymakers. In: *Climate Change 2021: The Physical Science Basis. Contribution of Working Group I to the Sixth Assessment Report of the Intergovernmental Panel on Climate Change* [Masson-Delmotte, V., P. Zhai, A. Pirani, S.L. Connors, C. Péan, S. Berger, N. Caud, Y. Chen, L. Goldfarb, M.I. Gomis, M. Huang, K. Leitzell, E. Lonnoy, J.B.R. Matthews, T.K. Maycock, T. Waterfield, O. Yelekçi, R. Yu, and B. Zhou (eds.)]. Cambridge University Press, Cambridge, United Kingdom, pp. 3–32, doi:10.1017/9781009157896.001.

Kidd, D. (2020). *A curriculum of hope: as rich in humanity as in knowledge*. Bancyfelin: Independent Thinking Press.

Klöckner, C. (2013) 'A comprehensive model of the psychology of environmental behaviour; a meta-analysis', *Global Environmental Change, Volume 23, Issue 5*, pp2013). https://doi.org/10.1016/j.gloenvcha.2013.05.014

Kolbert, E. (2014). *The Sixth Extinction: An Unnatural History*. London: Bloomsbury Publishing Plc.

Lambert, D., Roberts, M. and Waite, S. (2020). *The National Curriculum Outdoors*. London: Bloomsbury Publishing Plc.

Li, C.J. & Monroe, M.C. (2019). 'Exploring the essential psychological factors in fostering hope concerning climate change.' *Environmental Education Research, 25 (6), pp.936-954*.

Lloyd, D. and Paige, K. (2022). Learning science locally: Community gardens and our future. *Frontiers in Education*, 7.

Lumber, R., Richardson, M. and Sheffield, D. (2017). Beyond knowing nature: Contact, emotion, compassion, meaning, and beauty are pathways to nature connection. *PLOS ONE*, 12(5), p.e0177186.

Malone, K.A. and Waite, S. (2016). Student outcomes and natural schooling pathways from evidence to impact report 2016.

Mann, J., Gray, T., Truong, S., Brymer, E., Passy, R., Ho, S., Sahlberg, P., Ward, K., Bentsen, P., Curry, C. and Cowper, R. (2022). Getting Out of the Classroom and Into Nature: A Systematic Review of Nature-Specific Outdoor Learning on School Children's Learning and Development. *Frontiers in Public Health*, 10.

Marchant, E. *et al.* (2019). 'Curriculum-based outdoor learning for children aged 9-11: A qualitative analysis of pupils' and teachers' views', *PLOS ONE*, 14(5). doi:10.1371/journal.pone.0212242.

Morris, E.A. (2014). *Assessing & Developing Self-Esteem Ages 5-11*. Northampton: Loggerhead Publishing Ltd.

Ms. Lindsey's Book Nook (2021). *'Seed to Plant' ~ Plant Read Aloud ~ Plant Story time ~ Science Read Aloud ~ Garden Read Aloud*. YouTube. Available at: https://www.youtube.com/watch?v=AwBfDyllYDc [Accessed 19 Sep. 2022].

Natural England (2019). *Monitor of Engagement with the Natural Environment Children's Report (MENE) 2018-2019*. [online] GOV.UK. Available at: https://www.gov.uk/government/statistics/monitor-of-engagement-with-the-natural-environment-childrens-report-mene-2018-2019 [Accessed 8 Sep. 2023].

Natural History Museum (n.d.-a). *Biodiversity: Do your bit for nature*. [online] Discover Biodiversity | Natural History Museum. Available at: https://www.nhm.ac.uk/discover/biodiversity/act [Accessed 14 Sep. 2023].

Natural History Museum (n.d.-b). *How to grow a wildflower pot for pollinators*. [online] www.nhm.ac.uk. Available at: https://www.nhm.ac.uk/discover/how-to-grow-a-wildflower-pot-for-pollinators.html [Accessed 14 Sep. 2023].

Natural History Museum (n.d.-c). *National Education Nature Park and Climate Action Awards Scheme*. [online] www.nhm.ac.uk. Available at: https://www.nhm.ac.uk/about-us/national-impact/national-education-nature-park-and-climate-action-awards-scheme.html [Accessed 14 Sep. 2023].

Office for National Statistics (2020). *One in eight British households has no garden*. [online] www.ons.gov.uk. Available at: https://www.ons.gov.uk/economy/environmentalaccounts/articles/oneineightbritishhouseholdshasnogarden/2020-05-14 [Accessed 14 Sep. 2023].

Ofsted (2019). *Education Inspection Framework (EIF)*. [online] GOV.UK. Available at: https://www.gov.uk/government/publications/education-inspection-framework [Accessed 8 Sep. 2023].

Ojala, M. (2016). Preparing children for the emotional challenges of climate change. in Winograd K. (Ed.), *Education in times of environmental crises: Teaching children to be agents of change* (pp. 210–218). New York: Routledge.

Osterloff, E. (n.d.). *What is climate change and why does it matter?* [online] nhm.ac.uk. Available at: https://www.nhm.ac.uk/discover/what-is-climate-change-why-does-it-matter.html [Accessed 13 Sep. 2023].

Payton, L. (2021). *SAVING OUR SOILS: Healthy soils for our climate, nature and health*. [online] Available at: https://www.soilassociation.org/media/24941/saving-our-soils-report-dec21.pdf [Accessed 22 Sep. 2023].

Pfautsch, S., Wujeska-Klause, A. and Walters, J. (2022). Outdoor playgrounds and climate change: Importance of surface materials and shade to extend play time and prevent burn injuries. *Building and Environment*, 223.

Pörtner, H., et al. (2021). *IPBES-IPCC co-sponsored workshop report on biodiversity and climate change; IPBES and IPCC*. Paper presented at the IPBES-IPCC Co-sponsored workshop report on biodiversity and climate change; IPBES and IPCC.

Public Health England (2014). *From evidence into action: opportunities to protect and improve the nation's health*. [online] Available at: https://www.gov.uk/government/uploads/system/uploads/ attachment_data/file/366852/ PHE_Priorities.pdf

Roberts, M. (2021), Wild Tribe Subject Leader Award. [online] Available at: https://www.arena-schools.co.uk/wild-tribe/22153/wild-tribe-subject-leaders

Roberts, M. (2022), Wild Tribe Outdoor Learning: Wild Tribe Explorers Outdoor Learning Programme. [online] Available at: www.arena-schools/wildtribe.co.uk

Royal Horticultural Association (2022). Introducing the National Education Nature Park. [online] Available at: https://schoolgardening.rhs.org.uk/News/News-results/National/2022/November/National-Education-Nature-Park [Accessed 16 Oct. 2023]

Royal Horticultural Soceity (n.d.) *A Checklist of Potentially Harmful Plants / RHS Campaign for School Gardening*. Available at: https://schoolgardening.rhs.org.uk/resources/info-sheet/a-checklist-of-potentially-harmful-plants (Accessed: 16 October 2023).

Rickinson, M., Dillon, J., Teamey, K., Morris, M., Young Choi, M., Sanders, D. and Benefield, P. (2004). *A review of research on outdoor learning*. London: National Foundation for Educational Research and King's College London.

Rittel, H.W.J. and Webber, M.M. (1973). Dilemmas in a general theory of planning. *Policy Sciences*, 4(2), pp.155–169.

Rousell, D. and Cutter-Mackenzie-Knowles, A. (2019). A systematic review of climate change education: giving children and young people a 'voice' and a 'hand' in redressing climate change. *Children's Geographies*, 18(2), pp.191–208.

RSPB. (n.d.). *Big Schools' Birdwatch*. [online] Available at: https://www.rspb.org.uk/fun-and-learning/for-teachers/ schools-birdwatch/ [Accessed 19 Sep. 2023].

Save Our Wild Isles. (n.d.). *Save Our Wild Isles*. [online] Available at: https://www.saveourwildisles.org.uk/ [Accessed 6 Sep. 2023].

Scoffham, S. and Rawlinson, S. (2022). *Sustainability Education*. London: Bloomsbury Publishing.

Shulman, L. (1987). Knowledge and Teaching: Foundations of the New Reform. *Harvard Educational Review*, 57(1), pp.1–23.

Sobel, D. (2008a). *Childhood and nature: design principles for educators*. Portland: Stenhouse Publishers.

Tasquier, G., Pongiglione, F. and Levrini, O. (2014). Climate Change: An Educational Proposal Integrating the Physical and Social Sciences. *Procedia - Social and Behavioral Sciences*, 116, pp.820–825.

Teach the Future (2021). *Teach the Future's response to the Department for Education's draft Climate and Sustainability Strategy*. [online] Available at: https://uploads-ssl.webflow. com/5f8805cef8a604de754618bb/61a9143ab80963511809970c_20211201%20Teach%20the%20Future%20 response%20to%20DfE%20CCS%20Strategy.pdf [Accessed 7 Sep. 2023].

The Aquifer Partnership (n.d.). *The Aquifer Partnership*. [online] TAP. Available at: https://wearetap.org.uk/ [Accessed 14 Sep. 2023].

The Nature Friendly Schools (n.d.). *Home | Nature Friendly Schools*. [online] Available at: https://www.naturefriendlyschools.co.uk/ [Accessed 14 Sep. 2023].

Thunberg, G. (2019). *No one is too small to make a difference*. London: Penguin Books.

Thunberg, G. (2022). *The Climate Book*. London: Allen Lane.

Tippy Tap. (n.d.). *Tippy Tap*. [online] Available at: http://www.tippytap.org [Accessed 20 Sep. 2023].

Turns, A. (2022). *The UK wants to plant 120m trees a year by 2025. Can it be done?* [online] Positive News. Available at: https://www.positive.news/environment/can-the-uk-plant-120-million-trees-a-year/ [Accessed 19 Sep. 2023].

UNCED (1992). *United Nations Conference on Environment & Development: Agenda 21.* [online] United Nations. Available at: https://sustainabledevelopment.un.org/outcomedocuments/agenda21 [Accessed 6 Sep. 2023].

UNESCO (2020). Education for sustainable development: a roadmap. [online] Available at: https://unesdoc.unesco.org/ark:/48223/pf0000374802 [Accessed 7 Sep. 2023].

UNESCO (2021). *Berlin Declaration on Education for Sustainable Development Preamble.* [online] *en.unesco.org.* Available at: https://en.unesco.org/sites/default/files/esdfor2030-berlin-declaration-en.pdf [Accessed 7 Sep. 2023].

UNICEF (1992). *UN Convention on the Rights of the Child (UNCRC).* [online] UNICEF. Available at: https://www.unicef.org.uk/what-we-do/un-convention-child-rights/ [Accessed 11 Sep. 2023].

UNESCO (2015) Rethinking Education: Towards a Global Common Good. United Nations Educational, Scientific and Cultural Organization: Paris.

United Nations (n.d.). *Take Action for the Sustainable Development Goals.* [online] United Nations. Available at: https://www.un.org/sustainabledevelopment/sustainable-development-goals/ [Accessed 6 Sep. 2023].

Waite, S. (2020) 'Where are we going? international views on purposes, practices and barriers in school-based outdoor learning', *Education Sciences*, 10(11), p. 311. doi:10.3390/educsci10110311.

Waite, S., Davis, B. and Brown, K. (2006) Final report: *Five stories of outdoor learning from settings for 2-11 year olds in Devon*, July 2006, report for funding body EYDCP (zero14plus) and participants. Plymouth: Plymouth University.

Waite, S., Goodenough, A., Norris, V. & Puttick, N. (2016). From little acorns: environmental action as a source of ecological wellbeing, *Pastoral Care in Education: An International Journal of Personal, Social and Emotional Development.* 34 (1), 43-61. Available at: http://www.tandfonline.com/doi/full/10.1080/02643944.2015.1119879

Waite, S., Passy, R., Gilchrist, M., Hunt, A. and Blackwell, I. (2016). *Natural Connections Demonstration Project, 2012-2016: Final Report and Analysis of the Key Evaluation Questions - NECR215.* [online] Natural England. Available at: https://publications.naturalengland.org.uk/publication/6636651036540928 [Accessed 7 Sep. 2023].

Waite, S., Husain, F., Scandone, B., Forsyth, E. & Piggott, H. (2021). 'It's not for people like (them)': Structural and cultural barriers to children and young people engaging with nature outside schooling', *Journal of Adventure Education and Outdoor Learning*, 23(1), pp. 54–73. doi:10.1080/14729679.2021.1935286.

Waite, S. and Pratt, N. (2015). *Situated learning (Learning in Situ).* in J. D. Wright (Ed), International Encyclopaedia for the Social and Behavioural Sciences. 2nd ed. Oxford: Elsevier, p. 5012.

Warwick, P., Warwick, A. and Nash, K. (2017). *Towards a pedagogy of hope: Sustainability Education in the early years.* In V. Huggins & D. Evans (Eds.), Early Childhood Education and Care for Sustainability: International Perspectives (pp. 28-39). Abingdon, Oxon: Routledge.

White, M.P., Alcock, I., Grellier, J., Wheeler, B.W., Hartig, T., Warber, S.L., Bone, A., Depledge, M.H. and Fleming, L.E. (2019). Spending at least 120minutes a week in nature is associated with good health and wellbeing. *Scientific Reports*, [online] 9(1). Available at: https://www.nature.com/articles/s41598-019-44097-3 [Accessed 13 Sep. 2023]..

Whittaker, F. (2022). *Schools have surplus land 'the size of central London'.* [online] schoolsweek.co.uk. Available at: https://schoolsweek.co.uk/schools-have-surplus-land-the-size-of-central-london/ [Accessed 13 Sep. 2023].

Wilson, E.O. (1984). *Biophilia.* Cambridge, MA: Harvard University Press.

World Commission on Environment and Development (1897). *Our Common Future (The Brundtland Report).* Oxford: Oxford University Press.

World Wildlife Fund. (2017). *Protect Life on Earth.* [online] Available at: https://www.worldwildlife.org/videos/protect-life-on-earth--2 [Accessed 19 Sep. 2023].

Appendices

Appendix 1

Extracts from the government paper, published 21 April 2022: Sustainability and climate change: a strategy for the education and children's services systems

Available online at https://www.gov.uk/government/publications/sustainability-and-climate-change-strategy/sustainability-and-climate-change-a-strategy-for-the-education-and-childrens-services-systems

The following extracts, represented by quote marks, provide a rationale to underpin your development of a whole school policy to address SCCS through the curriculum. They set out why and how the government is planning this programme and the vital role that schools play. The extracts focus particularly on the parts of the document that are related to immediate actions that schools can take as the programme to 2030 unfolds. All author commentary is represented by italic text.

"Education

Through education we have the privilege to be able to engage directly with children and young people who:

- are passionate about the natural world

- want to do their best to protect it

- can influence their wider communities.

Through their learned and lived experiences from early years to further and higher education, we will provide opportunities to develop a broad knowledge and understanding of the importance of nature, sustainability and the causes and impact of climate change and to translate this knowledge into positive action and solutions.

In the UK, there are more than 16 million children, young people and adults in education. The enthusiasm of youth can inspire the whole of society to work together at the start of this crucial decade for the planet."

"Learning from and connecting with nature

The Economics of Biodiversity: The Dasgupta Review states that 'connection with nature declines in childhood to an overall low in the mid-teens. Creating an environment from an early age where we are able to connect to nature is essential for self-enforcement in protecting and valuing nature' (HM Treasury, 2021).

We will increase opportunities for all children and young people to:

- spend time in nature and learn more about it

- become actively involved in the improvement of their local environment.

We know that regular contact with green spaces can have a beneficial impact on children's physical and mental health. However, access to green space is not equal and we must do more to ensure that all children have opportunities to benefit from access to green space and build connections with nature."

The main focus of this book is adaptations to the curriculum and school grounds that enable learning about climate change as well as other interrelated sustainable development goals. Within the SCCS, these goals are mostly featured in:

"Action area 1: Climate education

We know that young people are eager to:

- create a greener, sustainable world

- tackle both the causes and impact of climate change.

We will empower all young people to be global citizens, through a:

- better understanding of climate change

- greater connection to nature.

Practical opportunities to participate in activities to increase climate resilience, reduce carbon impact and enhance biodiversity will enable children and young people to translate knowledge into positive action to improve their local communities, their country and the planet."

There are 3 main aspects to operationalising this ambition.

"1. Learning about the natural environment

Building on a foundation of fundamental numeracy, literacy and broad academic knowledge, all children learn about:

- nature

- the causes and impacts of climate change

- the importance of sustainability.

From birth to 5 years old, the early years foundation stage (EYFS) framework ensures that all children develop an understanding of the world and the natural environment.

As they progress through primary and secondary school, children and young people continue to build on this knowledge through science, geography and citizenship programmes within the national curriculum.

2. Support for teaching

World class teaching will ensure all children and young people get the best possible climate education. The schools white paper sets out how we will provide an excellent teacher for every child, by giving every teacher and school leader access to world class training and development opportunities, including the first-ever national professional qualification (NPQ) for early years professionals.

Through our engagement with the teachers and representative bodies, we have heard that more support in teaching about climate change and in navigating the many different resources available is also needed. Therefore, recognising the importance of building confidence as well as capability, we will provide additional support to teachers of all levels.

3. Learning in the natural environment

Education settings provide a wealth of learning opportunities, practical activities and clubs which allow children and young people to bring their learning to life. Children and young people may:

- take part in eco-clubs or vegetable growing

- be exposed to sustainable food choices, recycling, adaptation projects or weather and energy monitoring.

On top of the learning benefits, these activities can aid pastoral work in all educational settings. The physical and mental health benefits of time spent in nature can form part of targeted support to:

- improve engagement and attainment, including as part of wider packages of support for pupils with SEND

- give young people a sense of agency where anxiety stems from climate concerns.

The National Education Nature Park and Climate Action Award (*further details below*) will build on this excellent activity, ensuring all children and young people have opportunities to get practical experience and turn their knowledge into positive action. We will design these initiatives with inclusivity at their heart. We are committed to enabling those from disadvantaged backgrounds to access these opportunities."

Several support mechanisms for schools are already in place through the following initiatives.

"Initiatives to drive the strategy

We have carried out work with sector representatives and experts to develop initiatives that bring together activity to drive the strategic aims:

- increasing opportunities for climate education and access to nature

- driving opportunities to increase biodiversity and climate resilience

- co-ordinating and leading a whole-setting approach to climate change and sustainability."

"National Education Nature Park

By considering the whole physical education estate as a virtual National Education Nature Park, we have a unique opportunity to:

- deliver improvements in biodiversity

- contribute to the implementation of the nature recovery network

- play our part in halting nature's decline

- drive greater climate resilience.

The National Education Nature Park will:

- engage children and young people with the natural world

- directly involve them in measuring and improving biodiversity in their nursery, school, college or university

- help reinforce their connection with nature.

This will help connect and amplify the excellent work already happening in this area through many national and local stakeholders and community groups.

As their work starts to have an impact, the young participants will upload their progress on the park's digital mapping services. They will be able to:

- see how the park is 'growing', while increasing their knowledge of species

- develop important skills, such as biodiversity mapping, data collection and analysis.

The nature park's online hub will enable the sharing of best practice across the education estate. With time, it will go beyond biodiversity to show impact and climate resilience, including flood, overheating and air quality status of the estate.

A number of universities will support the launch of the National Education Nature Park by:

- acting as champions of nature and biodiversity for local education settings and wider communities

- providing opportunities to share their expertise and natural environment

- supporting other education settings in developing and delivering a better environment for future generations."

"Climate Action Award

A Climate Action Award will complement classroom learning and allow us to celebrate and recognise education providers, children and young people for:

- developing their connection with nature

- making a real contribution to establishing a sustainable future for us all.

The award will provide a structured route through existing awards in this area, such as the John Muir Award, Duke of Edinburgh's Award, Junior Forester Award and others.

Participation will enable children and young people to acquire credits towards the prestigious Climate Action Award. This will be recognised and valued as supporting progression to employment and further study.

We want all young people to feel connected to their local environment and see improvements in biodiversity. In the design and implementation of the park and award we will take steps to drive participation among more disadvantaged children and young people. We will also ensure all children and young people, whether they live in an urban area or rural one, have opportunities to feel empowered through practical positive action.

The National Education Nature Park and Climate Action Award:

- are being designed and developed with young people, sector representatives and stakeholder organisations

- will build on the learning and experience of the Year of Green Action (Department for Environment, 2019).

They will feed in ideas for making the initiatives as inclusive, impactful, and engaging as possible and will also consider the best names for the programmes. The park and award will be launched in autumn 2022."

If you would like to receive further updates about the development and delivery of the strategy, you can sign up to their mailing list. Signing up to receive the **Climate In Education Snapshot,** *visiting www.educationnaturepark.org.uk and getting their email alerts or following hashtag* **#ClimateInEducation** *on social media will keep you posted on the latest news!*

Engaging with these initiatives will help you keep abreast of the resources and advice available to schools over the coming years and promote leadership.

Appendix 2

Case Study (in full)

Droxford Junior School: School grounds transformation

"Back in 2019 one could have been forgiven for thinking our school grounds were idyllic; a lovely, green rectangular field, surrounded by trees, nestled in hills within the South Downs National Park. And it was! I have taught in many places, including the East End of London so I know how fortunate we were breathing in this vista! Lovely to look at but a dynamic space for learning, play and a haven for an abundance of wildlife? No, not really, especially as it was seasonally 'out of bounds'.

Growing into my new role as headteacher, I embarked on an experiential journey to rectify this situation, planning to develop the school grounds to enhance learning and play, promoting wellbeing as well as providing space for the community.

The initial proposal was aligned with our 'Skills For Learning' ethos; the characteristics and habits of mind we endeavour to nurture, teach and reward in school. The proposal also linked to our School Vision: 'Equipping our children with the knowledge and skills to navigate life successfully'. One of the most important elements to support the proposal was to engage someone experienced in landscape strategy. This is where the knowledge, skill and experience of a Landscape Architect, Catherine Eldred of Hampshire Local Authority, was key. Catherine Eldred collated and provided chapters for the planning process, having the area pictured, described, scored and perceptions noted. The school raised funds to pay for this support by crowdfunding through the Aviva Community fund.

You can read more about this process here: https://www.droxfordjunior.co.uk/page/?title=Outdoor+Learning+and+Play&pid=26

The most important component for making any project like this successful is consensus and near-universal support – a buy-in for all. Gaining the 'voice' of the whole school community, not just listening but hearing the voices, views and opinions of others was vital; as important as securing funding in order to be confident of genuine support for the project. Initial workshops and consultations to share views and suggestions on the proposed plans were undertaken to collect to the opinions and viewpoints of the children, staff, parents and governors of the school. Gathering the advice of many could be seen as an onerous task, but actually the process is a powerful one as it enables everyone to realise that they are not a lone voice, that many others agree with the ambition and through this the vision gains momentum. Developing a collaborative long-term strategic plan for the grounds means everyone in the school has a shared positive vision. It is then not a design created by a consultant but is created by the entire school and everyone within it. It is made all the better for being the sum of so many parts.

<u>So what did the process look like in practice?</u>

The children were provided with numbered photographs of spaces around the school site. We asked every child to rate these spaces and score them out of 10, with 1 being the lowest, and to also, where possible, qualify their scores. Staff, governors and parents were also involved in this process.

I'd like to stress to all, that being honest and authentic is vital for helping to gain support. If our grounds were already perfect, we wouldn't have been going through the process, which was very a reflective one. It was important not to take things personally and for everyone to understand others' perceptions and views was key. After all, we need opinions and viewpoints – without them we couldn't make positive changes.

Here are some examples:

- **Location 4** is called the Main Playground or the Three Thirds Playground. It is used for a range of PE activities including netball and basketball and for free play at break and lunchtimes. It feels 'busy, crowded, sometimes unsafe, cramped, tight, squashed, fun but overwhelming'. Quality of the space is considered good as it has a variety of uses, but also there is not much room when it is zoned as 2/3 for sport and 1/3 for playing. Use of the area was rated as 9/10 as used every day for a wide variety of physical activity that is appreciated by many.

- **Location 5** has a number of names including the Wildlife Area, the Nature Reserve, the corner of the field and also the Dump. The area is thought of as a space where bug hotels are made, bonfires are lit, and 'as a peaceful space to sit in in the summer'. The area is described as feeling 'unloved, messy, unused, dull, uncomfortable and unhappy for wildlife', by some, but also 'creative, peaceful, relaxing calm and spacious' by others.

- **Location 6** is known by everyone as The Field. It is used for PE, clubs, football, summer play and 'running around in'. To those within school it feels 'free, spacious, fun, great, joyful, with interesting side bits, but also with a sense of emptiness'. Use of the field is currently seasonal as those at the school mentioned it gets very muddy in the winter, and so becomes out of bounds for much of the year. Quality was rated as 8.5/10.

We asked the school community – 'Where do we want to be?'

Our vision of 'Equipping children with the knowledge and skills to navigate life successfully' coupled with me being very hands on with this vision, meant that I had to let go a bit but still share my thoughts. We carried out a workshop with the school community, asking 'We would like our school grounds to be a place where...?'. Some examples of the answers are below:

- We can enjoy ourselves

- Where the outside continues to enable physical wellbeing

- Where creativity can be encouraged

- First impressions and spending time outside school is welcoming to all

- Where we respect our environment

- Where the grounds support and promote learning as much as the buildings do

- Where the grounds support emotional and mental wellbeing

- There is year-round accessibility to all the grounds

- Where the whole school community get to enjoy a closeness to the natural world

- We feel safe and secure.

We asked 'What key spaces would we like in our grounds?'

This second workshop was an opportunity to focus on previous conversations about the current site, and to discuss the future developments.

We learnt a lot from the answers to both questions and although the suggestions for a future swimming pool or 'Go Ape' were kindly dismissed, the children's comments were mature and insightful – sound bites I still use today!

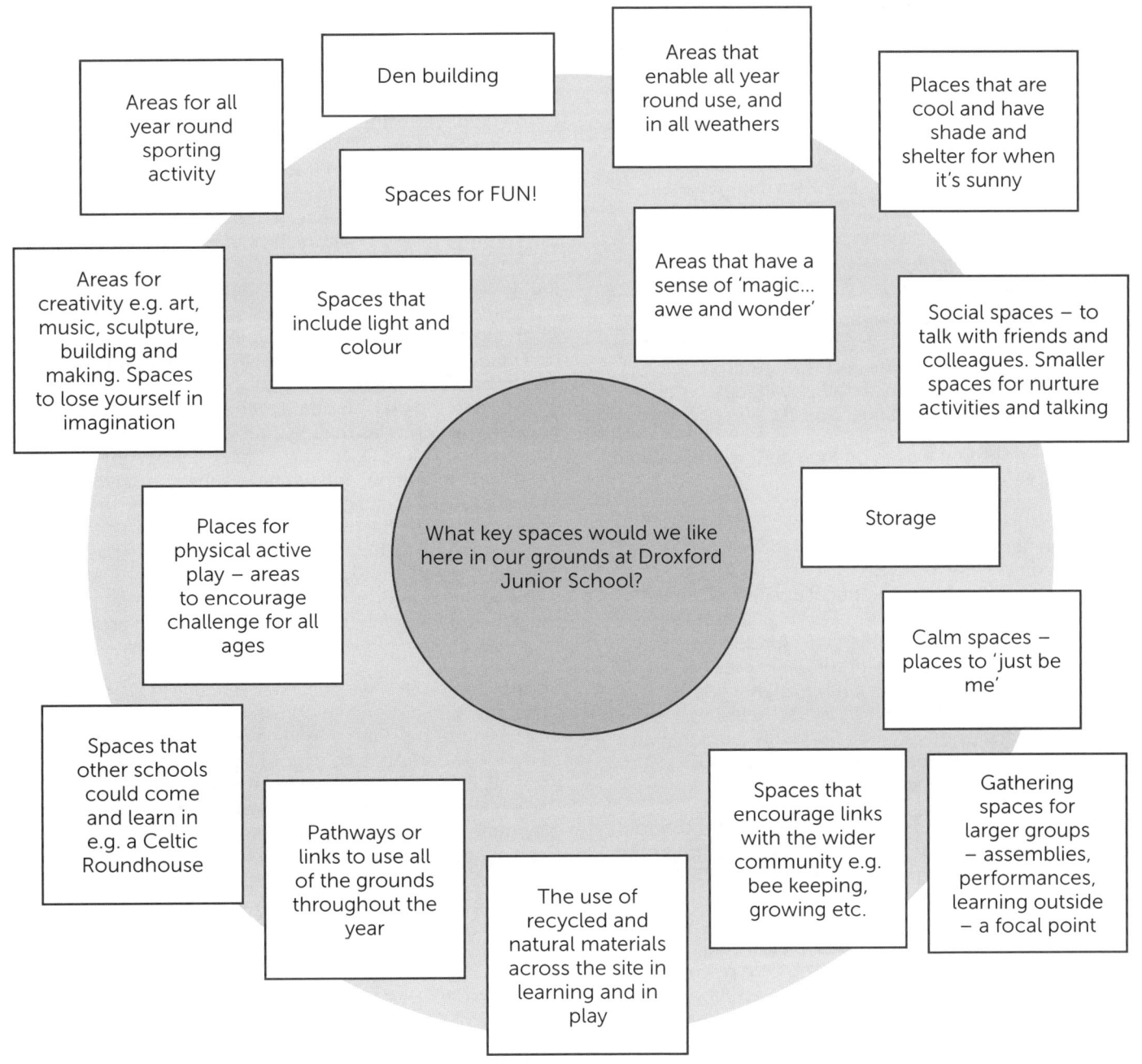

With this achieved, a plan of the school site was drawn up and elements of all of the above were discussed with the school grounds team from the LA: Catherine Eldred; the South Downs National Park team (Amanda Elmes and Jonathan Dean), local horticulturalists (Angela & Andy Ward from the Butterfly and Bee Plant Nursery in Cheriton, Hampshire) and school staff, to finalise the plan. We took into consideration the aspect of the field and soil type, where trees were, site-lines and the best positions for any new installations. We also considered what we didn't want to change, including sporting events; PE curriculum and Friends of Droxford School (FODS) activities and promotions.

Key Principles:

- To respect the natural world and the environment around us here in Droxford.

- To encourage our values of **Respect, Kindness and Responsibility** through every part of school life, including the use of our grounds and the wider environment.

- We also believe children need to be given opportunities to be Creative, Tenacious, Collaborative and Curious. This strategy plan will ensure children have the opportunity to achieve this. Each 'Area' has these Key Principles linked to them.

From strategy into action

Each zone of the strategy had its own mini action plan – often in my head or on scraps of paper (not healthy or ideal I know). Leading change and staff getting used to and trusting change helps with the buy in, but it is also important to share that 'change' doesn't always work.

The first development was to change culture. This would be (and still is) a recurring theme.

The first culture change was to get the children and staff on the field all year round and in all weathers, if they wanted to. This meant developing and agreeing strategies to overcome the practical challenges, like welly storage and muddy shoes. This was a relatively inexpensive improvement, which was funded by a sponsored fun-run by the children. Using creative ways to fundraise for projects, with explanations in newsletters, assemblies and emails also helps with raising the profile of the project. Suddenly, our smallish playground is freer of energy, potential flash-points and frustrations with lack of space.

Word began to spread to everyone linked to the school, that we had big plans for learning, play and getting more wildlife into the school grounds. By now, our FODS were fully on board with supporting the Landscape Strategy for a year or two and the children were extremely positive because they were allowed on the field all year round, whatever the weather. With the help of parent donations and FODS, we purchased Outlast Blocks (wooden version of LEGO®), tarpaulins, ropes, hay bales and a variety of poles etc. We also scavenged unwanted PE equipment to establish areas in the grounds where children could be creative with loose parts play, to den-build and construct. Now, children are occupied when off the main field and to this day den-building is the most important and fun activity, noticeably for children with SEND, in line with our 'Skills for Learning' – Creativity, Tenacity, Collaboration and Creativity.

However, bigger plans require more money. After extensive and exhaustive form-filling for grants from a variety of sources with no success and with an ever-increasing sense of deflation, it was decided that we should approach the South Downs National Park Community Infrastructure Levy (CIL) for funding for the whole project. Having a clear plan was key to the grant funding. This was where our holistic vision came into its own. We could show the plan, we could show anyone and everyone how the entire school were part of this, and we were vocal about how developing our grounds wasn't purely for educational purposes but to improve health and wellbeing in a wide variety of ways, for as many parts of our school and its wider community as possible. Although initially unsuccessful, during the first COVID lockdown we heard that we'd been successful, and we could go ahead with all of the zones and complete to a high standard!

The next phase was to develop a play area at the bottom of the field. We wanted something natural and not too manufactured – stepping stones and logs to jump across, to look like fallen trees, in keeping with the natural surroundings. With further enquiry, we found we had a parent who happened to be a tree surgeon. In the summer of 2021, natural timber was brought in and positioned in a design the children felt would be a great (but lower) version of 'Go Ape'. Tactically, by being lower than 600mm we didn't need any special insurance. As an extra bonus the logs acted like seats and we had inadvertently created our own natural outdoor classroom! The site of the log play-trail is in the shade of tall trees, which is perfect for the summer but not so good in the winter as the logs get (slightly) slippery. We did our own risk assessment, which the children played a big hand in writing, and we have subsequently chain-sawed grooves to aid grip. When the logs were first in position children fell off of them left, right and centre! Nothing major, just bumps and bruises. Agility has improved greatly now and they fly across them! There is the added bonus of risk-benefit analysis being part of the process too, with the children learning to assess and take risks as part of their daily decision making.

Prior to the project we had established 'Dig Days' at the school, which took place on Saturdays in the autumn and spring terms. One thing I have always endeavoured to do is get parents in to school as much as possible. After all, it's their school as much as anyone else's. Parents were able to bring children if they wanted to, to play quite happily in the grounds with friends and in the fresh air and we fed everyone at lunchtime with pizza and salad and cake, all made in the school kitchen, with tea, coffee and biscuits on tap. If you have a parent with a mini-digger, parents in trades and parents who are keen gardeners, feed them pizza and cake!

Over the years, we have created an allotment (with a shed) and maintained it by repairing raised beds, weeding and planting and painting out-buildings in the school grounds. Our parents have been incredible! With a new sense of purpose, the Dig Days really ramped up in importance and engagement. The South Downs National Park Rangers joined in, teaching us about our locality and a volunteer has helped the children to make wooden bird boxes. Over the course of six months and two Dig Days we created and planted a nature corridor and completely repurposed an unloved corner of the school field. The nature corridor has since been planted with six new trees, especially selected for shade, pollinators, fruit and berries and aesthetic design. These have all been under planted with pollinator-friendly plants and spring bulbs, which in turn helps increase and improve the biodiversity within the grounds, an aim from the very beginning. The wildlife area has now been planted with wildflowers and plants for pollinators all year round. There are three bug hotels and a stag beetle hotel plus new posts included to provide homes for solitary bees. The area is access-friendly and has a chip bark path that loops all the way through it. What once was an area full of brambles, nettles, burnt Christmas trees and rubble, is now an engaging area for quiet, contemplative thoughts and nature. It is a lovely place to sit and relax; an area that will grow and change over time.

Between the wildlife area and the end of the nature corridor is a beautiful, bespoke hexagonal hut open on two sides, with benches inside for 30 children. It isn't anything fancy, no thatch or electrics, no door or windows, just a shelter we call 'Cedar Base' (it has a cedar wood roof) inspired by a building at the Gilbert White Field Study Centre. Having a contractor who is fantastic at working in natural materials, to personal sketches and to tight budgets, and who totally fulfilled the brief (and even provided wood stumps which we placed in a circle for our firepit) helped. A shout-out to Andy Smith (Rhythm and Hues). During lockdown we purchased 12 sturdy picnic benches from our local secondary school. These immense pieces of wooden furniture will last for years and were made by students in the DT department during lockdown. We now have enough seating for over 70 children spread around the periphery of the field. Some seating has been re-purposed by the children in the den building area with a tarpaulin draped over them. The perfect den; a table and bedsheet!

But we didn't stop there! With the funding achieved and available, we endeavoured to complete as much of the plan as possible. Shade in school grounds is often a lacking resource and in our increasingly hot summer months it was becoming more vital at lunchtimes, when all of us should be able to be outside, enjoying some fresh air and space. Andy Smith helped again with the creation of a terraced seating area, adjacent to the playground, which provides shade for moments of free play, an area for pupils and parents to watch sporting activity on the wider field and a designated theatre space for performances or where lessons can be held. Additional planting here also gives a clearly defined edge to the playground. The former failing wooden sleepers have been replaced to provide flower beds that once again will fill with colour and scent and will attract additional wildlife.

What action was taken next, in order to make a difference to attracting wildlife in and around the school site?

We really wanted to know the difference our project would make, attracting wildlife in and around the school site. So, at the start of the planting process, before the professional horticulturalists were brought in, an ecology study of the site was commissioned to provide accurate baseline data. We had two very detailed reports of what was seen growing in the grounds and what wildlife was seen, at certain times across the year, which forms part of a longitudinal study.

We engaged with Learning through Landscapes (LTL) and the Hampshire and Isle of Wight Wildlife Trust (HIOWWT). We now have an eco-committee, with members taken from across the school (including the children). As part of our HIOWWT Wilder Committee actions we have ten target insects we are aiming to see in the grounds, that we haven't seen before. To highlight this target across the wider school community, 'wanted' posters for the insects have been placed in the newly developed and planted garden in the courtyard. Our children have been involved in pollinator studies, although we have only just started on this journey, and much is dependent on seasonal plants and their habitats.

It is important the children meet people who bring with them a wealth of experience, fresh enthusiasm and knowledge. Engaging with experts from LTL and the HIOWWT has not only enriched the school curriculum but has potentially opened up a new world of opportunity for the pupils, offering new directions in life that they weren't aware were open to them.

Stopping there would be easy wouldn't it? Surveying all that we have done, and being proud that our grounds offer so much more now is only half the story. Being an Ambassador School for the National Park (the only one in Hampshire) and a hub school for the Wessex Rivers Trust is important to us because it raises the profile of the school in developing the delivery of outdoor education not just in our school, but other settings too; to overcome any barriers and feel confident in going outside more often. Our site has been developed with this in mind. A Landscape Strategy is a living document that should and will change over time, and part of the culture change was to address a slight nervousness about taking learning outside the classroom. Many teachers have this nervousness as they strive to meet targets or achieve goals and often a nice warm classroom with books and technology on-hand seems like the easiest and perhaps safest place to achieve these. But consider, who from their childhood remembers the indoor maths lesson? Rather, who remembers the time outside to explore the grounds, to use all senses, to listen and hear sounds to inspire a story or to write poetry? We knew that developing the confidence of our staff to see the opportunities that school grounds provided, both as a source of learning and for health and wellbeing, was something not just talked about by us. So, we opened up our learning in an Outdoor Education Collaboration CPD day aiming to raise the profile of outdoor learning across the county (and wider). We welcomed as many as wanted to come along, providing the opportunity to learn from a fascinating and experienced group of guest speakers with practical workshops shared with more than 40 schools from as far as Guernsey, and for all ages of pupils from infants to sixth-form college.

As part of the South Downs National Park, we thought that our grounds should be available to those beyond Droxford (in particular for those who might never venture even a few miles from their homes). So, we invited and hosted 45 Year 4 children from an urban school to come to us for an overnight residential in May 2022, accompanied by some of their teaching staff. We provided tents for them to sleep overnight in, which was an experience most had never had before. They visited the River Meon. They cooked and ate marshmallows around the firepit and were woken by early sunrise and birdsong. Every child asked to return again and their school has already booked up again! So, that was the next part of the plan; that our school grounds should be available to others from more urban contexts, to enable these children and teachers to have a connection with the natural world too.

We are now at a point where I know we have achieved a lot in 18 months, with much pre-planning. The profile of outdoor education at Droxford has been raised by its inclusion in our School Improvement Plan. Staff have been tasked with developing their subjects with more focus on outdoor learning. My headteacher performance management is linked to increasing the profile of outdoor learning and I want to run another Landscape Strategy Plan this summer to find out where are we now, including finding out what differences we have made and what our new cohort of children want for their future. The school targets and school council have this noted as their key task.

There is still much to do. The site looks wonderful and when the plants begin to bloom this spring I am confident we will see some of our 'target' insects identified on the Wilder Committee 'wanted' posters. But we are not a finished article. We want to see more wildlife in school and learn how we attract it, but we also need to evaluate whether what are we doing will increase happiness and wellbeing. How do we know our children are healthier and happier? *Is it impacting on engagement with learning and raised attainment?* What will the ecology study say next time? What can we do next? How can this be made sustainable?

The constant cycle of planning through the seasons never fails to inspire. One person cannot do it all. I am so grateful for support in the school community and all mentioned in this case study. Not only are we strategically planning for our children, but we are involved in supporting a wider cohort of children. Long may this continue."

Matthew Dampier, headteacher, Droxford Junior School March 2023

Appendix 3

Sithney Community Primary School, Cornwall, grounds development plan for outdoor learning 2022 – 2027

Wild Tribe Outdoor Learning

Curriculum Intent and Drivers/7 Themes of an effective Outdoor Learning Environment

- The ecological impact of current and future uses of areas and how this can be minimised
- Development plan ideas
- Links to curriculum Intent and drivers
- Links to the 7 Themes of an effective Outdoor Learning Environment

			PERSEVERANCE			RESPECT		VALUES
CURRICULUM INTENT	The National and Early Years Foundation Stage Curriculum in England	Commitment to the UN Global Action on Climate Change	KNOW YOURSELF	MAKE EXCELLENT PROGRESS	ENJOY BEING ACTIVE	CELEBRATE DIFFERENCE AND DIVERSITY	MAKE A DIFFERENCE	VISION
			Resilient Individuals	**Respectful Communicators**	**Healthy Advocates**	**World Citizens**	**Environmental Ambassadors**	DRIVERS
			Children to show resilience; to be able to approach challenges positively in all areas of life. To be independent and reflective learners, utilising these skills to enable a positive future.	Children to be effective communicators, interacting with confidence in a variety of situations. To be respectful and respond to others in a meaningful way.	Children to live healthy and active lifestyles. To promote positive, physical and mental health that will help to provide a strong foundation for their future.	Children to have a sense of their own belonging within the local, wider and global communities. To show a deep respect for the diversity of our world.	Children to have an experience-rich practical understanding of the environment. To show they care about the management and sustainability of our planet.	
			STAND ON YOUR OWN TWO FEET	WORK WELL TOGETHER	DREAM BIG	EXPLORE AND CONTRIBUTE	EMBRACE THE OUTDOORS	VISION
			KINDNESS			ASPIRATION		VALUES

Sithney CP School Curriculum Intent and Drivers

7 Themes of an effective Outdoor Learning Environment

- Adventure
- Fantasy and Imagination
- Animal Allies
- Maps and Paths
- Special Places
- Small Worlds
- Hunting and Gathering

ZONE 1 Active Zone Sports / Athletic Field and Active Playground		
WHAT IS THERE?		ACTIONS
NATURAL MATERIALS	• Grass Open Land • Newly planted trees • Wildflowers • Fungi • Sports / athletics field • Active playground **Learning Focus:** This Zone is where the children demonstrate: • The key values of perseverance, kindness, resilience and respect. Children learn how to: • Know themselves, make excellent progress, enjoy being active, stand on their own two feet and work well together.	1. Promote use of the centre of field and playground for sport and athletics. 2. Provide maintenance to newly planted trees including clearing area around base of trees **(Animal Allies)**. 3. Complete wildflower survey and promote wildflower growth on hedgerows and periphery of field **(Animal Allies)**. 4. Allow natural growth and die-back of fungi in the centre of the field during Autumn months **(Animal Allies)**.
ECOLOGICAL IMPACT These actions will help to enhance the **biodiversity** of Zone 1 whilst supporting the children to **Make Excellent Progress** and **Explore and Contribute**.	**CURRENT** • Maintenance of current tree species in field. **POTENTIAL FUTURE** • Development of planters for sensory plants. • Development of Reading area. • Creation of recycling / compost station.	1. Maintenance of trees, wildflowers and fungi to promote biodiversity. 2. Promote future wildflower growth. 3. Felling of tree 673 due to Ash Dieback. 4. Weed control / Woodchip mulch at base of newly planted trees.
FUTURE PLANS		ACTIONS
1. Sensory planters		1. Development of planters for sensory plants in playground **(Hunting and Gathering)**.
2. Outside Reading Area		2. Development of outside Reading area in playground where oil tank used to be located **(Fantasy and Imagination)**.
3. Recycling Station		3. Creation of recycling / compost station in playground **(Hunting and Gathering)**.

Immediate, Medium and Long-term Development plan of the school's grounds

STAGE 1 – IMMEDIATE 6-12 months	
DEVELOPMENT PLAN	**CHALLENGES**
Zone 1 1. Maintenance of trees, wildflowers and fungi to promote biodiversity. 2. Promote future wildflower growth. 3. Felling of tree 673 due to Ash Dieback. 4. Weed control / Woodchip mulch at base of newly planted trees. **Zone 2** 5. Maintenance of trees – removal of dead branches / ivy **(Animal Allies)**. 6. Maintenance of outdoor learning space / orchard with regular weed control / removal **(Animal Allies)**. 7. Development of Leylandii space to better act / link in with clamber as an adventure zone **(Animal Allies)**. **Zone 3** 8. Maintenance of weeds and brambles alongside regular pruning and cutting **(Small Worlds)**. **Zone 4** 9. Promote as a quiet / reflective space for children where they can get to know themselves, dream big **(Fantasy and Imagination / Special Places)**. 10. Low maintenance to promote wild nature of area **(Special Places)**.	**Zone 1** • Time / cost to mulch newly planted tree. • Time / cost to promote additional wildflower growth. • Cost of tree surgeon to remove tree 673. **Zone 2** • Time to provide regular maintenance. **Zone 3** • Time to provide regular maintenance. **Zone 4** • Clarity of purpose for spaces among all staff / cascading of information contained in development plan.
SOLUTIONS	
• Use of Sports Premium budget to fund activities involving children and promoting physical and healthy learning experiences. • Incorporate routine maintenance of Zone 2 into outdoor learning activities. • Walk through learning zones with key staff to discuss aims and actions included in development plan to better enable information to be cascaded and incorporated in future outdoor learning planning / resourcing.	
BENEFITS	
• Dangerous trees felled and at-risk trees / areas maintained to promote better growth and sustainability. • Clarity of vision and purpose of outdoor learning plan.	

STAGE 2 – MEDIUM TERM 1-2 years	
DEVELOPMENT PLAN	**CHALLENGES**

DEVELOPMENT PLAN	CHALLENGES
Zone 1 1. Development of planters for sensory plants in playground **(Hunting and Gathering)**. 2. Development of outside Reading area in playground where oil tank used to be located **(Fantasy and Imagination)**. 3. Creation of recycling / compost station in playground **(Hunting and Gathering)**. **Zone 2** 1. Turn cultivated bed area into pond area to further enhance biodiversity **(Animal Allies)**. 2. Creation of a path (Roman Road) to link quiet area to rest of field **(Maps and Paths)**. 3. Development of quiet area (Willow dome) in western corner of field **(Special Places)**. 4. Development of Adventure Zone to incorporating: Clearing under hawthorn and putting down wood chip **(Adventure)**. 5. Purchase / revamp mud kitchen in orchard area **(Small Worlds)**. 6. Installing knot board in outdoor learning area to promote knot tying and den building **(Adventure)**. **Zone 3** 1. Install raised beds to make planting simpler and more straightforward for all year groups. This will make more efficient use of growing vegetables and reduce constant need for maintenance **(Hunting and Gathering / Animal Allies)**. 2. Tidy / sort / organise outdoor classroom, tool shed and greenhouse so that they are well resourced and functional working spaces to ensure a whole class can access this learning space at any one time **(Small Worlds)**. **Zone 4** 1. Secret Path – further clear and develop secret path to ensure that it is not too overgrown and enables access whilst still being hidden and wild **(Special Places)**. 2. Install hammock between Oak trees so that the children can relax and enjoy nature whilst being inspired by nature and dreaming big **(Fantasy and Imagination)**.	**Zone 1** • Time / cost to create / purchase planters. • Time / cost to resource outside reading / writing area. • Time / cost of resourcing recycling station. **Zone 2** • Time / cost to resource materials needed to create pond. • Time / cost to resource materials for path and willow dome. • Time / cost to remove material and purchase mud kitchen, knot board etc. **Zone 3** • Time / cost to install raised beds and organise space. **Zone 4** • Maintenance and weed control to ensure secret path is cleared and managed. • Purchase of hammock for reflection space.

SOLUTIONS

• Estimated costs of materials etc. needed to achieve medium term plans would be approximately £1000 – £1500. Sports Premium funding could be used to support these costs as it is providing children with additional access to healthy and active learning opportunities.

• School and Eco Council alongside FOSS could be utilised to help research / plan and fund some of proposed developments.

• Incorporate routine maintenance of Zone 2 into outdoor learning activities.

• Walk through learning zones with key staff to discuss aims and actions included in development plan to better enable information to be cascaded and incorporated in future outdoor learning planning / resourcing.

BENEFITS

• Learning environment will be further enhanced and enable children to engage in more purposeful cross curricular learning opportunities. Many of the outcomes directly support the School Development Plan.

STAGE 3 – LONG TERM 1-5 years	
DEVELOPMENT PLAN	**CHALLENGES**
Zone 1 1. Development of dirt track / mounds with cleared soil from other excavation areas. **Zone 2** 2. Utilising existing tree stumps by creating totem poles to represent school values and leadership traits linking to world citizens curriculum values **(Fantasy and Imagination)**. Attach a variety of hooks and fixings to enable rope work and creation of assault courses **(Adventure)**. **Zone 3** 3. Enhance community engagement and collaboration by starting parent and baby group **(Small Worlds)**. **Zone 4** 4. Install boat to enable children to have a special place where they can develop fantasy and imagination **(Fantasy and Imagination)**.	**Zone 1** • Time / plan to make this a meaningful exercise to reuse excavated soil from development of pond and quiet space with willow dome. **Zone 2** • Time / expertise / cost to liaise with local artist to carve totem poles. **Zone 3** • Time / staffing to initiate and start group. **Zone 4** • Time / research / cost of suitable object / installation to capture the fantasy / imagination of children.

SOLUTIONS

- Opportunities to work with CAST as possible organisation to support an artist to work with the school on the totem pole project. Costs for such a project should be between £1000 - £2000.
- Community artist and parent and children group might qualify for a Lottery Awards for All grant to support costs.
- Competition within school for children to design / select an installation for the secret garden which would capture their fantasy / imagination. Local charities such as HRCST could be approached to donate an old boat beyond repair which would still function as an installation. Equally, children may decide on a different installation and the costs of these could be incorporated into a Lottery grant to promote health and wellbeing.

BENEFITS

- Greater links with the community will strengthen the school's links and promote the school as a positive place for children to enrol.
- Children having greater involvement in the design of learning space and decision making process will empower them and directly links to the school's curriculum drivers.

Appendix 4

Chapter 3: Brief audit tool for sustainability status of grounds

Circle the number which shows how much you disagree or agree with each statement: 1 means strongly disagree, 5 means strongly agree, i.e. you feel you have these features securely in place. If you feel you are on your way towards these features, choose a number between 2 to 4 depending on how far along you are. Strike through any statements that you do not wish/feel able to pursue, justifying your view. Add other features. Reflect on your scores and use this tool to help formulate an action plan.

	Our School Grounds' sustainability features	Disagree → Agree				
1	We have considered biodiversity and climate change factors in the development of our grounds	1	2	3	4	5
2	We have made use of expertise to advise us in actions for climate change and sustainability improvements in our grounds	1	2	3	4	5
3	Pupils are involved in the development and maintenance of grounds	1	2	3	4	5
4	Our local community is involved in the development and maintenance of grounds	1	2	3	4	5
5	We have a wide variety of different features and zones in our grounds	1	2	3	4	5
6	We have mature native trees in our grounds	1	2	3	4	5
7	We have planted native trees to grow on	1	2	3	4	5
8	The trees are located where they (will) provide useful shade for hotter parts of the site	1	2	3	4	5
9	We have a meadow zone, where we let grass and wildflowers grow tall	1	2	3	4	5
10	We grow pollinator plants that encourage insects to visit our grounds	1	2	3	4	5
11	We have areas where deadwood and fallen leaves provide habitat	1	2	3	4	5
12	We have a water feature to support aquatic life	1	2	3	4	5
13	We have a water feature to gather rainwater	1	2	3	4	5
14	We have stormwater retention or channelling systems for extreme rain conditions	1	2	3	4	5
15	We have lower-level native shrub and hedging to provide shade cover and corridors for wildlife	1	2	3	4	5
16	We have a forest garden for sustainable fruit and other edible plant production with trees, bushes and ground cover plants	1	2	3	4	5
17	We have a garden/ raised beds/ containers to grow vegetable and fruit	1	2	3	4	5
18	We have pathways that allow us to access wild areas without disturbing them	1	2	3	4	5
19	Where we can, we reduce, reuse and recycle materials to create our grounds or source them locally	1	2	3	4	5
20	We have introduced manmade wildlife friendly features such as bird boxes and insect hotels	1	2	3	4	5
21	We use some form of renewable energy	1	2	3	4	5
22	We have reduced food waste and compost organic matter on site	1	2	3	4	5
23	We encourage cycling or walking to school	1	2	3	4	5
24	We use different zones in the school grounds to maximise teaching and learning opportunities	1	2	3	4	5
25	We have improved the biodiversity in our site	1	2	3	4	5
26	Our school grounds are excellent for teaching and learning outside	1	2	3	4	5
27	Our school grounds are excellent for meeting climate change and sustainability challenges	1	2	3	4	5

This brief audit tool is based on Chapter 3 and its sources. For a complete audit tool, contact Learning through Landscapes for their help.

Appendix 5

Chapter 3: Envisioning tool with children

Ask pupils to draw a picture of their school grounds with all the features that they would like to see in it. Then swap their pictures with a partner.

Partners should make a list in two columns of all the things that would benefit people and all the things that will benefit nature. Which is longer? Why do they think this is the case? Were some things good for both people and nature? Discuss whether humans and nature are distinct – can we be separated from nature? Are we being fair to non-human species with whom we share the planet?

Give a copy of the picture of the Swedish school garden on the linked website to the pairs. ◑ Can they make a list with three columns that shows things that are good for people, good for non-human species, or good for both?

Discuss our interdependence with nature and how looking after other species is also good for us.

Ask them to draw another picture and show what things they would want to include to help reduce the impact of a changing climate on us and other species.

Use this information to provide a pupil perspective on what school grounds improvements to work on.

Appendix 6

Chapter 3: List of possible UK trees for shade

Alder *(Alnus glutinosa)*
Beech *(Fagus sylvatica)**
Downy birch *(Betula pubescens)*
Elder *(Sambucus nigra)**
Rowan *(Sorbus aucuparia)*
Sessile oak *(Quercus petraea)*
Hornbeam *(Carpinus betulus)*
Ash *(Fraxinus excelsior)*
Blackthorn *(Prunus spinosa)**
Silver birch *(Betula pendula)*
Hazel *(Corylus avellana)**
Field maple *(Acer campestre)**
English or Pedunculate oak *(Quercus robur)*
Holly *(Ilex aquifolium)**
Hawthorn *(Crataegus monogyna)**

*These are also suitable for hedging.

Appendix 7

Chapter 3: List of suitable plants

RHS Guidance on features to help increase wildlife in the school garden can be found at:
https://schoolgardening.rhs.org.uk/Resources/Info-Sheet/Encouraging-wildlife-in-your-school-garden

They suggest good general pollinator plants are:

Cosmos bipinnatus (Cosmos) – Sow in spring

Borago officinalis (Borage) – Sow in spring and summer

Calendula officinalis (English Marigold) – Sow in autumn or spring

Phacelia tanacetifolia (Phacelia) – Sow in late spring and summer

Centaurea cyanus (Cornflower) – Sow in autumn or spring

Eschscholzia californica (Californian poppy) – Sow in spring

Helianthus annuus (Sunflower) – Sow in late spring

Lavatera trimestris (Annual Mallow) – Sow in late spring

Limnanthes douglasii (Poached Egg Plant) – Sow in autumn or spring

Nigella damascena (Love-in-a-mist) – Sow in autumn or spring

If you want to attract **Bees** to the school site, plant:

Allium species
Bird's foot trefoil (*Lotus corniculatus*)
Clovers (*Trifolium* species, for example red clover and white clover)
Cornflower (*Centaurea cyanus*)
Cotoneaster species
Cranesbill (*Geranium* species)
Crocus species
Devil's bit scabious (*Succisa pratensis*)
Firethorn (*Pyracantha* cultivars)
Golden rod (*Solidago* species)
Heliotrope (*Heliotropium* cultivars)
Hemp agrimony (*Eupatorium cannabinum*)
Honesty (*Lunaria annua*)
Lavender (*Lavandula* species)
Love-in-a-mist (*Nigella damascena*)
Marjoram (*Origanum* species)
Michaelmas daisies (*Aster* species)
Poached egg plant (*Limnanthes douglasii*)
Types of single-flowered rose species (*Rosa* species)
Rosemary (*Rosmarinus officinalis*)
Rosebay willowherb (*Chamaenerion angustifolium*)
Scabious (*Scabiosa* species)
Snowdrops (*Galanthus* species)
Sunflower (*Helianthus annuus*)
Teasel (*Dipsacus fullonum*)
Thyme (*Thymus* species)

Butterflies will be especially drawn to:

Alyssum (*Lobularia maritima*)
Aubretia (*Aubrieta deltoidea*)
Blackberry (*Rubus fruticosus*)
Brazilian vervain (*Verbena bonariensis*)
Bugle (*Ajuga reptans*)
Candytuft (*Iberis amara*)
Catmint (*Nepeta x faassenii*)
Dandelion (*Taraxacum officinale*)
Globe thistle (*Echinops* species)
Heather (*Calluna vulgaris*)
Hebe species
Ice plant (*Sedum spectabile*) – not dark red cultivars
Knapweeds (*Centaurea* species)
Marigold (*Calendula officinalis*)
Mignonette (*Reseda odorata*)
Slender vervain (*Verbena rigida*)
Thistle (*Cirsium* species and *Carduus* species)

Plants that are particularly popular with **Birds** according to the RHS include:

Berberis species
Birchleaf viburnum (*Viburnum betulifolium*)
Crab apples (*Malus* species)
Elderberry (*Sambucus nigra*)
Field scabious (*Knautia arvensis*)
Guelder rose (*Viburnum opulus*)
Hawthorn (*Crataegus monogyna*)
Ivy (*Hedera helix*)
Some species of rose (*Rosa rugosa* and *Rosa moyesii*)
Wild roses (*Rosa canina* and *R. rubiginosa*).

Mixes of wildflowers will attract many different types of pollinators and their seeds can be directly scattered into grassy areas. More success may be achieved if they are grown on first and transplanted. Remember that most wild plants do not like soil that is too rich in nutrients and will compete with grass better where the soil is poor and grass thinner.

Appendix 8

Chapter 4: Progression 3

Questions and answers

What does sustainability mean?

Doing things differently so resources last longer and we can protect planet Earth.

What is biodiversity?

It is all the various kinds of plants, animals, fungi, and microorganisms that make up our natural world.

What is rewilding?

It is turning an area to a wilder or more natural state which will help animals and insects return.

How can we save water?

Using a water butt or a container to collect water will save water that we can reuse.

What is a bug hotel?

It is a place where animals can make their home. They are often made from wood, mud and straw or grass.

Can you name an item that you could recycle?

You can recycle items made of plastic, metal and paper.

What would you like to protect on planet Earth?

Trees, oceans, rainforests, animals, plants are all things that we need to protect on planet Earth.

Why do we need trees?

They give us oxygen and they breathe in carbon dioxide, helping to keep the air clean.

Appendix 9

Chapter 4: Progression 6

Examples of how children can take action at their school

Rewilding: Make pinecone bird feeders to encourage more birds around your school.

Recycling: Run a competition where students pick up litter and recycle on their way to and from school each day. Students who take part are entered into a prize draw to win cinema vouchers. (This is inspired by the fantastic ideas of Treviglas School in Cornwall).

Recycling: Make your own eco brick by filling a plastic bottle with clean, dry used plastic. Eco bricking is a simple way to take personal responsibility for plastic use by keeping it out of industry and the biosphere.

Reduce: Think of ways to reduce the use of paper.

Reduce: Create 'free for all' stalls for fruit and vegetable snacks that are not eaten by the end of the school day to reduce food waste.

Reuse: Promote using a reusable bottle for drinking water each day.

Climate change: Taking conservation action and showing others how to do the same, e.g. planting trees, beach clean-ups, restoring community gardens, etc.

Climate change: Walk or cycle to school where possible.

Climate change: Organise or attend an event, e.g. a pop-up thrift shop at school, a healthy eating event in a local park, litter picks, nature festivals, etc.

Climate change: Take part in Earth Hour by not using electricity for the time spent completing the Earthrise Leadership Award.

Appendix 10

Chapter 5: Progression 1

Examples for 'big question' cards

- What different habitats or ecosystems are in the outdoor area? E.g. flowerbeds, vegetable plots, hedgerows, fields, paved areas, etc.
- What evidence might you find of animals living in a location? E.g. webs, snail trails, walkways, footprints, spoors, feathers, nests, holes, nibbled leaves, etc.
- What do habitats or ecosystems provide? E.g. places to forage and feed, safe places to breed and shelter, etc.
- What do animals need to survive? E.g. space to breed and shelter
- Habitats can be quite specific, so which outdoor area provides the most suitable habitat for, for example, snails?
- Where are invertebrates most likely to be found? Why is this?

Appendix 11

Chapter 5: Progression 2 and 3

Examples for 'big question' cards

- What host plants might attract insects, birds or animals? Why is this important?
- What different ecosystems are in the outdoor area? E.g. flowerbeds, vegetable plot, hedgerow, field, paved area, etc.
- What do these ecosystems provide? E.g. places to forage and feed, safe places to breed and shelter, etc.
- What do plants need to grow?
- Why are plants important?
- What host plants might attract insects, birds or animals?
- Why is it important for plants to attract insects?
- How might plants be classified into groups?
- What features of plants help with classification?

Appendix 12

Chapter 5: Progression 3

Possible questions to think about when considering why some areas of the school grounds are more biodiverse than others

- Is there any evidence/sign of animals (including invertebrates) in the area? Why are they here? What are the animals' characteristics? What are the implications of these characteristics on this type of location?

- Is there any evidence of change caused by human impact, such as rubbish, worn areas, planned gardening or horticulture? What impact might this have had on habitats or biodiversity?

- Is there any evidence of the impact of quality, moisture content or texture of the soil? What might the cause be, e.g. too much or too little rain?

- Are the plants in the outdoor area wild plants, native plants or have they come from another part of the world, such as invasive species? What impact might this difference have?

- Are all the plants the same? Are there different coloured flowers in particular areas? Might colour make a difference in attracting insects? What impact would this have on attracting other creatures?

- Are there different trees? What observable features do they have, such as leaf shape or colour? Might different trees attract specific wildlife or insects?

- How have plants adapted to the different conditions? E.g. shallow roots in gravel, moss growing in moist areas, plants layering to reach the sunshine, etc.

- Are there any differences that might impact the quality of areas, such as changes to the soil?

- Is there any evidence of the impact of too much sunlight or lack of sunlight? E.g. specific plants found in boggy areas, gravel areas, etc.

- Is there any evidence of weeds thriving or not thriving in barren areas?

- Is there any evidence of previously thriving areas that are now saturated (too much rain) or dry (no rain), which might indicate climate impact?

- Is there any evidence of lichens or moss, which may indicate air quality?

Appendix 13

Chapter 5: Progression 3

Possible differences in planting in areas marked red or green on the RAG rating

- A 'no mow' policy for grass areas has created wild areas.

- A variety of plants has been carefully selected for year-round flowering, offering a mix of colours, shapes, sizes, and fragrances, alternatively, native trees, shrubs, and flower meadows have been planted to attract pollinating insects.

- The soil has been dug and left fallow (unseeded).

- A living nature corridor has been created by planting along a border.

- A dead hedge has created opportunities for feeding, reproduction or shelter.

Appendix 14

Chapter 5: Progression 4

Posing questions to investigate and devising fair tests

Some factors of growth that could be tested include:

- having lots of sunlight (or not)

- watering daily (or not)

- providing food such as fertilizer (or not)

- varying the soil quality (or not)

- adding heat by creating a cold frame (or not)

- growing indoors or outdoors.

Example fair test of 'having lots of sunlight (or not)':

Soak broad beans in water, place them on a damp paper towel or cotton wool and place them in a sandwich bag, before choosing differing growing areas, e.g. hanging in a window or in a dark cupboard. Ensure all other growing conditions are the same. Use this test to find out what happens when the beans have lots of sunlight (by the window) and no sunlight (in the cupboard).

If you want to read more from these authors...

The National Curriculum Outdoors series
By Deborah Lambert, Michelle Roberts and Sue Waite

Aimed at improving outside-the-classroom learning for children from Year 1 to Year 6, this series of photocopiable lesson plans effectively link curriculum objectives for each primary subject with motivating outdoor learning. This enables pupils to experience the health and wellbeing benefits of learning in the natural environment, and offers progression in knowledge and skills across the key stages.

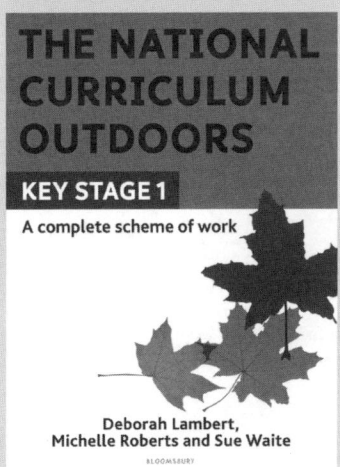

'These books are a timely and welcome addition to help primary teachers grow their confidence and competence to undertake great teaching - outdoors!'

Juliet Robertson, former headteacher and education consultant specialising in outdoor learning, @CreativeSTAR

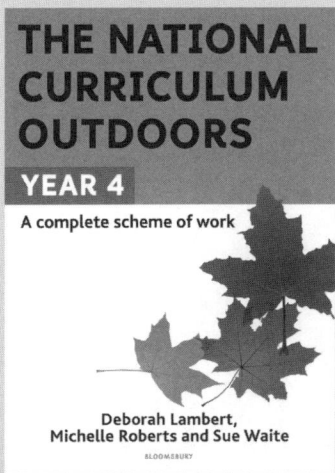

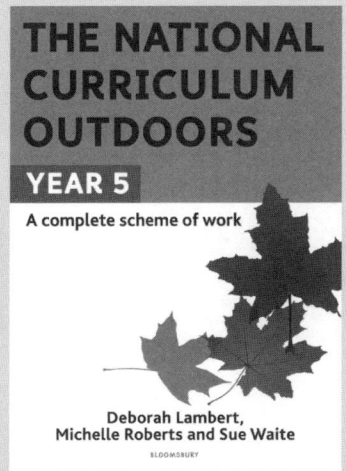

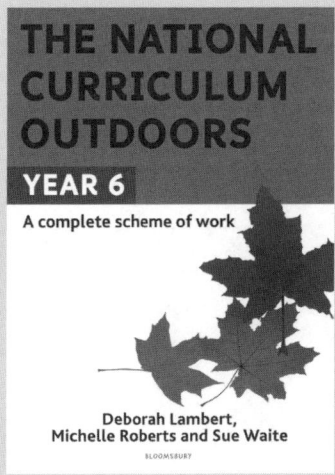